UNDERSTANDING
CLONING

UNDERSTANDING CLONING

FROM THE EDITORS OF *SCIENTIFIC AMERICAN*

Compiled and with introductions by

Sandy Fritz

Foreword by William A. Haseltine

a Byron Preiss Book

WARNER BOOKS

An AOL Time Warner Company

Copyright © 2002 by Scientific American, Inc.,
and Byron Preiss Visual Publications, Inc.
All rights reserved.

The essays in this book first appeared in the pages and on the Web site of *Scientific American*, as follows: "The Regeneration of Potato Plants from Leaf-Cell Protoplasts" May 1982; "Plantibodies" November 1997; "Cloning Noah's Ark" November 2000; "A Clone in Sheep's Clothing" March 1997; "Cloning Hits the Big Time" September 1997; "Cloning for Medicine" December 1998; "Transgenic Livestock as Drug Factories" January 1997; "Mother Nature's Menders" Special Issue 2000; "Xenotransplantation" July 1997; "I, Clone" Special Issue 2000; "Synthetic Self-Replicating Molecules" June 1994; "Go Forth and Replicate" August 2001.

Warner Books, Inc., 1271 Avenue of the Americas, New York, NY 10020
Visit our Web site at www.twbookmark.com.

For information on Time Warner Trade Publishing's online publishing program, visit www.ipublish.com.

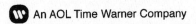 An AOL Time Warner Company

Printed in the United States of America
First Printing: March 2002
10 9 8 7 6 5 4 3 2 1

ISBN: 0-446-67874-0
Library of Congress Control Number: 2001093910

Cover design by J. Vita
Book design by Gilda Hannah

Contents

Foreword vii
BY WILLIAM A. HASELTINE

Introduction 1
BY SANDY FRITZ

**The Regeneration of Potato Plants
from Leaf-Cell Protoplasts 7**
BY JAMES F. SHEPARD

Plantibodies 17
BY W. WAYT GIBBS

Cloning Noah's Ark 23
BY ROBERT P. LANZA, BETSY L. DRESSER
AND PHILIP DAMIANI

A Clone in Sheep's Clothing 37
BY TIM BEARDSLEY

Cloning Hits the Big Time 41
BY TIM BEARDSLEY

Cloning for Medicine 45
BY IAN WILMUT

Transgenic Livestock as Drug Factories 57
BY WILLIAM H. VELANDER, HENRYK LUBON
AND WILLIAM N. DROHAN

Mother Nature's Menders 69
BY MIKE MAY

Xenotransplantation 79
BY ROBERT P. LANZA, DAVID K. C. COOPER,
AND WILLIAM L. CHICK

I, Clone 89
BY RONALD M. GREEN

Synthetic Self-Replicating Molecules 99
BY JULIUS REBEK, JR.

Go Forth and Replicate 111
BY MOSHE SIPPER AND JAMES A. REGGIA

**Remarks by the President on Stem Cell
Research 123**
BY PRESIDENT GEORGE W. BUSH

Conclusion 129

Index 130

had cloned an adult sheep
ediately captivated the publ
some cautious researchers
ther the success might not
the pan. After all, the Scot
ters had to try 277 times be
ceeded in producing the clor
ly. Unless the efficiency of
ning process could be great

Foreword

William A. Haseltine

*T*o Clone: To reproduce, by human hand, identical copies
of a living organism.

Cloning, whether reproductive or therapeutic, is a
controversial topic today, and will be the subject of intensive
public debate for decades to come. Questions about this new
technology number far more than answers. Is reproductive
cloning a fundamental departure for our species, or is it just
another means, such as in vitro fertilization, to produce a baby?
Is cloning another human being technically too risky to con-
template, or can the risk be made equivalent to, or reduced
below, that of normal reproduction? Today, one in five concep-
tions ends in miscarriage and about one in ten babies is born
with serious developmental problems. Will therapeutic
cloning, the use of tissues derived from newly created
embryos, ever find a place in medicine, and is it ethical? Must
new embryos be created especially for therapeutic purposes, or
will all-purpose embryonic stem cells suffice?

Is it possible to imagine a time when medicine will move from a focus on regeneration of injured organs, tissues and cells, to rejuvenation—making the body young again? Will it be possible through rejuvenative medicine, enabled by stem cell cloning, to extend human life beyond its upper six score year limit? And, if it is possible, should we do it?

Some may be surprised at what has already been achieved. Cloning of cattle will soon be routine. Inherited properties of laboratory and of farm animals can be changed *ad perpetuitatem* by implantation of new genes. Genetically altered plants and animals can produce new medicines. Others may produce new industrial chemicals. One chapter describes how extinct species may be resurrected. Another focuses on what I would term rejuvenative medicine: how our own cells, made young again by cloning, may be used to rebuild organs, tissues and cells injured by trauma, damaged by disease, or worn by time.

The chapter, *I Clone*, addresses human cloning—the re-creation of near-identical humans. It outlines some of the perils, prospects and pitfalls. Important issues of identity, ethics and morality, although not settled, are at least raised therein.

The questions and debate that surround cloning will not end soon. This debate takes us to the farthest scientific frontiers and involves some of the deepest aspects of what we consider to be human identity and mortality.

My hope is that this book will be read by all those who are seriously interested in this debate. *Understanding Cloning* provides an essential background with regard to both the history and the current reality of this field. Before we can go forward, we must know where we have been. We must be able to distinguish fact from science fiction.

Understanding Cloning is a book that is in fact easy to understand. It is an important and timely contribution, a "must read" for all who wish to join the great debate about the present and future of cloning.

Introduction

Sandy Fritz

S ome 70 years ago, cloning was a word used mainly by horticulturists. Today, the word stands for a science that represents a nexus, a key turning point brimming with possibilities, that could well shape the future of our world.

How does one handle a science that allows for the creation of new life forms? Cloning, which can create an entire plant from a single cell, or alter the very living physiology of an animal, holds enormous promise for the human race. This science permits modified animals, that never walked the Earth, to be conjured, and perhaps, in the not too distant future, animals that once walked the Earth may return. It also implies enormous responsibility. Until now, natural forces alone have shaped the genetic codes of flora and fauna. Now, the ability to consciously shape and change the living things in our world is in our hands.

"Awkward but ingenious" best describes animal cloning in the early 21st century. The recipe is simple: Start with one female donor egg, say from a cow. Remove cell nuclei. Replace with a cell from, say a genetically-altered, prize-winning bull.

Return egg to cow womb. Allow to gestate. If the newly transferred cell was taken from embryonic fibroblast cells, the birth ratio stands at about one live birth for every ten tries. If the cell happened to be from an adult animal, such as in the case of Dolly, the World's First Cloned-From-An-Adult Sheep, the success rate drops to one live birth in 277 tries.

The varied success rate of these experiments reflects a science in its infancy. Although a barrier, it has not hampered creative thinkers who ponder and experiment with cloning's potential applications, especially in the field of medicine. Forward-thinking researchers are snipping a little animal DNA here and stitching a little human DNA there to produce transgenic animals. Some transgenic animals are like living drug factories, with the added human elements in their genetic makeup appearing as antibodies and other medicines in their milk.

Other transgenic animals can be created with organs far less likely to be rejected by the human body. The most visible strides are being made within the arena of combining human genes with animal genes, because the potential for discovery and breakthrough is closest to the surface.

In other directions cloning may be the key to creating and sustaining one of the most amazing cells in the human body, the stem cell. Stem cells are like blank slates that, when activated by certain chemical signals, can turn into any cell in the body. They power the growth of the human embryo, while in the adult body they are harbored and hoarded within bone marrow and selective caches. The idea of cloning stem cells, then directing their adaptive nature to repair a damaged heart or to regenerate failing hearing, could profoundly change the nature of medicine.

Behind the miracle of being able to manipulate a lifeform's genes lurks the need to know how it works, what sort of instructions are carried from cell to cell when the call to grow and multiply appears. Specialists producing simple molecules that clone themselves have first hand experience with the mind-boggling complexity of producing even the most simple structures.

As we braid new strands of DNA into an animal's genetic code, do we really know what fuels an organism's urge to extend? Can we pinpoint which factors marshal certain cues to quicken certain development, then vanish? When the mysteries behind the science stand out, the current state of cloning seems humble indeed. But it is a beginning, and a beginning with great promise.

Cloning has botanical roots and has been used by people for centuries to isolate the most desirable crops. Plants themselves have employed cloning for far longer, sometimes as their sole reproductive strategy, sometimes in tandem with sexual reproductive strategies.

A new wrinkle in the old fabric appeared at the close of the 20th century when researchers began to study small terminal leaves, the point where a plant grows the most quickly. When the leaf's cell walls were dissolved and treated through a series of steps, the researchers were able to coax a whole plant from a mass of cells.

This type of cloning, with leaf cell protoplasts, is common in the tobacco growing regions of the world.

The Regeneration of Potato Plants from Leaf-Cell Protoplasts

James F. Shepard

I n plants as in animals, variation from one individual to another is generally brought about through the shuffling of genes in sexual reproduction. The adaptive value of such variation is suggested by the elaborate reproductive organs of the flowering plants and by the equally elaborate strategies that have evolved to facilitate cross-pollination. Even so, not all flowering plants are totally dependent on sexual methods of reproduction. Some species have evolved an additional asexual reproductive capability, embodied in tissues that are anatomically distinct from the flower parts. The asexual, or vegetative, reproductive organ of such a plant can be an underground stem, a modified root or even a leaf that has the ability to develop into a complete plant. Whatever the origin of the vegetative tissue, asexual reproduction tends to preserve the phenotype, or physical characteristics, of the parent in its offspring.

Individual organisms that arise asexually from the somatic, or body, cells of the parent rather than from the specialized sexual cells are called clones, and the propagation of species by such methods is called cloning. I shall describe here a new experimental approach to cloning; perhaps surprisingly, it yields potentially useful forms of phenotypic variation in the regenerated plants. The technique, called protoplast cloning, has been developed primarily for potato plants, but in principle it could be applied to a wide range of crops.

A clone is often presumed to be a carbon-copy replica of the parent: "a chip off the old block." At first such a conclusion seems reasonable. In nature clones derived from the same parent are in many cases very similar to one another, and so they might be expected to have functionally equivalent genomes. This commonly held idea, however, is a misconception.

The term clone is derived from the Greek "kl'vn," meaning a slip, or twig, suitable for plant propagation. In 1903 Herbert J. Webber of the U.S. Department of Agriculture proposed that the transliterated word "clon" be adopted to designate those plants that are propagated asexually and (in his words) "are simply parts of the same individual." Soon afterward "clon" was altered in spelling to clone, and since then usage of the term has expanded to the point where it is now applied to all asexually reproduced forms of life. Indeed, the word has been applied even to the reproduction of DNA, the genetic material; thus one now speaks somewhat loosely of the cloning of genes in bacteria.

Today most scientific reference works continue to define a clone in the more restricted sense as an individual organism derived asexually from a single cell through mitosis, the form of cell division in which the daughter cells retain the same number of chromosomes as the parent cell has. This definition is not meant to imply genetic homogeneity in the resulting population. Clones from a single parent need not be identical in appearance or have exactly the same genetic composition. Indeed, in some species a clone can differ noticeably from its parent; it follows that the genome of a clone, like that of an organism reproduced sexually, must be equipped in some way to generate variation. This second source of variation has significantly extended the range of options available to the plant breeder in his quest for improved plant varieties.

Several crop species are propagated asexually to preserve essential varietal characteristics, either because the plants are sexually infertile or because their genomes are too complex for their phenotype to remain uniform in sexually reproduced progeny. In these species, clones that differ in some obvious way from the parent sometimes appear. Such divergent individuals are called somatic variants, bud sports or simply sports; they result from permanent genetic changes in specialized cells in the rapidly dividing cells at the tip of a growing stem, branch or root that generate all or part of the new plant. Many important varieties of clonally propagated crop plants have arisen from such vegetative mutations. Familiar examples are the pink grapefruit, the navel orange, the nectarine and several varieties of potato. In other plants, for instance the sweet potato, sports appear with a frequency that can be as high as 2 percent; as a result, the maintenance of varietal purity through conventional cloning is a continuing problem.

In recent years plant cloning has been refined to the point where a single cell removed from the body of a plant can be cultured and then induced to regenerate a complete individual. This developmental potential of single cells, a property known to cell biologists as totipotency, was first suggested by the results of experiments with cultured carrot cells done by Frederick C. Steward of Cornell University about 25 years ago. By about 1965 totipotency was firmly established for comparable tissue cultures of tobacco and other plant species; thereafter it was demonstrated for somatic cells and sexual cells isolated directly from numerous plants.

A major conceptual advance in this area was reported in 1971 by Itaru Takebe and his colleagues in Japan. They employed a combination of two enzymes, pectinase and cellulase, to dissociate tobacco leaves into living but wall-less plant cells called protoplasts. The isolated protoplasts were cultured in a medium that promoted growth and cell division. In the last step of their procedure the masses of cultured cells, called calluses, were induced to regenerate small shoots, which eventually grew into entire plants.

In 1973 my colleagues and I began working with leaf-cell protoplasts of tobacco. We soon found phenotypic differences between some members of the protoplast-derived clonal populations. In the absence of any mutagenic treatment variegated color patterns appeared in the leaves of roughly one out of every 250 regenerated plants. In subsequent sexual cross-breeding experiments the altered characteristics were usually passed to the offspring only through the maternal cells of the plant. Such a pattern of inheritance suggests that the mutations were in some genetic element outside the cell nucleus.

These findings implied that the tobacco protoclones with a normal chromosome number were not all identical. Accordingly we were encouraged to believe that if similar techniques could be applied to a crop species that naturally undergoes frequent somatic mutation, potentially useful forms of genetic change might ensue. To test this hypothesis we chose a common commercial variety of potato named the Russet Burbank.

The origin of the Russet Burbank potato dates back to 1875, when the botanist Luther Burbank selected a seedling from the progeny of a single potato fruit, or berry. The seedling was asexually propagated through the planting of tubers (the familiar edible portion of the potato plant, which is not a root but a modified underground stem); repeated cloning of this plant type eventually gave rise to the variety designated Burbank. Soon after the turn of the century a sport was selected from the Burbank line that produced tubers with a russet, or reddish brown, skin. The variant was given the name Russet Burbank; today it is the most widely grown variety of potato in the U.S., accounting for almost 40 percent of the total production.

Burbank's achievement in providing the basis for a new

potato cultivar from just a few seeds of one plant can be better appreciated if one considers that in the past 50 years more than 20 million potato seedlings have been evaluated by plant breeders in the U.S., but none has led to a commercial variety as successful as the Russet Burbank sport. It seems reasonable to predict, therefore, that given such a sound genetic foundation on which to build, definite opportunities exist for developing improved versions of the cultivar through sexual means. The mere selection of naturally occurring sports is neither an efficient approach nor a realistic one, however, because the frequency of somatic change for individual traits is low, and few if any spontaneous improvements in disease resistance, for example, have been recognized.

As an alternative we set out to measure the frequency of somatic variation in clonal populations derived from single protoplasts; the task would then become one of determining whether any advantage could be gained over the method of selecting sports at the plant level. To test this approach we first had to develop techniques for isolating the protoplasts of potato cells and for regenerating complete plants from them. In 1977 we published a preliminary description of a successful procedure, and since then we have refined the techniques and extended them to other potato cultivars.

Potato plants are susceptible to late blight, a disease caused by the fungus *Phytophthora infestans*. This disease, which caused the Irish "potato famine" of the 1840's, is still a major threat to commercial potato farming. Genes for resistance to the fungus have been introduced into some cultivars from wild relatives of the potato through conventional breeding techniques, but the fungus itself is highly variable, and strains emerge that can quickly overcome a simply inherited gene for resistance to the disease. Longer-lasting and broader-spectrum protection has sometimes been achieved by amalgamating several genes into what is termed a multigenic form of resistance.

Of all the traits measured in protoclonal populations so far, tuber yield is of particular interest. It is almost always desirable to increase the yield of a crop if commensurate increases are not also necessary in the energy, labor and other inputs of production. In potato farming yield means the weight of tubers produced per hectare of land, with the stipulation that the tubers be of a quality suitable for their intended use. Tuber quality in turn is defined by factors such as uniformity of shape and size, susceptibility to disease, storage characteristics and suitability for processing. It is essential that there be no compromise in such qualitative traits when increases in tuber weight are sought. In other circumstances the elimination of a single flaw in tuber quality can be as significant economically as increasing the weight of the tubers.

Statistical analysis of data from a 1979 field plot in North Dakota revealed that none of the 65 protoclones evaluated was superior to the parental clone in total tuber weight. The tuber yield of one protoclone, designated No. 307, did increase by 25 percent, but inconsistencies between different plantings rendered the increase insignificant. In 1980 follow-up field plots at the same location showed that once again protoclone No. 307 exceeded the yield of the parental Russet Burbank clone, but again the differences were not statistically significant. More years of experience are needed to determine whether No. 307 or any other protoclone offers a genuine yield advantage over the parent.

What is clear is that superior protoclonal performance at one geographic site is no guarantee of a similar advantage at another site. For example, in 1980 the same protoclones we tested in North Dakota were planted at a field plot in Colorado by Richard Zink of Colorado State University. He found that protoclone No. 307 did less well there than the parent, whereas other clones that did poorly in North Dakota equaled or exceeded the yield of the parental Russet Burbank clone. Site specificity has been a common phenomenon for the 65 protoclones over the past five years, and therefore it appears to be predictable. If it is, the Russet Burbank protoclones would resemble those varieties of potato in which the yield advantage among competitive lines is commonly realized only in certain geographic or environmental regions.

But there is a wide range of possibilities for creating new hybrid lines by fusing the protoplasts of genetically different plants prior to regeneration. Preliminary experiments along these lines are being done by our group and others. For example, we have succeeded in fusing a leaf-cell protoplast from a potato plant with one from a tomato plant. The goal of this particular effort is to introduce specific genes (for instance, those for disease resistance) from the tomato to the potato. Such an achievement would make possible the shuffling of genes between plant species that are now sexually incompatible, thereby increasing the size of the germ-plasm pool available to the plant breeder.

The ability to harness the growth of a plant to produce certain medical remedies exists today. Unfortunately, the fit between the medicinal molecules made by a plant and the human body is not quite perfect—yet.

This bold new world of plants that make human proteins and antibodies could move high-tech plant cloning into overdrive.

Plantibodies

W. Wayt Gibbs

own a country road in southern Wisconsin lies a corn-field with ears of gold. The kernels growing on these few acres could be worth millions—not to grocers or ranchers but to drug companies. This corn is no Silver Queen, bred for sweetness, but a strain genetically engineered by Agracetus in Middleton, Wis., to secrete human antibodies. In 1997, a pharmaceutical partner of Agracetus's plans to begin injecting cancer patients with doses of up to 250 milligrams of antibodies purified from mutant corn seeds. If the treatment works as intended, the antibodies will stick to tumor cells and deliver radioisotopes to kill them.

Using antibodies as drugs is not new, but manufacturing them in plants is, and the technique could be a real boon to the many biotechnology firms that have spent years and hundreds of millions of dollars trying to bring these promising medicines to market. So far most have failed, for two reasons.

First, many early antibody drugs either did not work or provoked severe allergic reactions. They were not human but mouse antibodies produced in vats of cloned mouse cells. In recent years, geneticists have bred cell lines that churn out antibodies that are mostly or completely human. These chimeras seem to work better: in July 1997 one made by IDEC Pharmaceuticals passed scientific review by the Food and Drug Administration. The compound, a treatment for non-Hodgkin's lymphoma, will be only the third therapeutic antibody to go on sale in the U.S.

The new drug may be effective, but it will not be cheap; cost is the second barrier these medicines face. Cloned animal cells make inefficient factories: 10,000 liters of them eke out only a kilogram or two of usable antibodies. So some antibody therapies, which typically require a gram or more of drug for each patient, may cost more than insurance companies will cover. Low yields also raise the expense and risk of developing antibody drugs.

This, Agracetus scientist Vikram M. Paradkar says, is where "plantibodies" come in. By transplanting a human gene into corn reproductive cells and adding other DNA that cranks up the cells' production of the foreign protein, Agracetus has created a strain that it claims yields about 1.5 kilograms of pharmaceutical-quality antibodies per acre of corn. "We could grow enough antibodies to supply the entire U.S. market for our cancer drug—tens of thousands of patients—on just 30 acres," Paradkar predicts. The development process takes about a year longer in plants than in mammal cells, he concedes. "But start-up costs are far lower, and in full-scale production we can make proteins for orders of magnitude at less cost," he adds.

Plantibodies might reduce another risk as well. The billions of cells in fermentation tanks can catch human diseases; plants don't. So although Agracetus must ensure that its plantibodies are free from pesticides and other kinds of contaminants, it can forgo expensive screening for viruses and bacterial toxins.

Corn is not the only crop that can mimic human cells. Agracetus is also cultivating soybeans that contain human antibodies against herpes simplex virus 2, a culprit in venereal disease, in the hope of producing a drug cheap enough to add to contraceptives. One company is testing an anti-tooth-decay mouthwash made with antibodies extracted from transgenic tobacco plants while another, CropTech in Blacksburg, Va., has modified tobacco to manufacture an enzyme called glucocerebrosidase in its leaves. People with Gaucher's disease pay up to $160,000 a year for a supply of this crucial protein, which their bodies cannot make.

"It's rather astounding how accurately transgenic plants can translate the subtle signals that control human protein processing," says CropTech founder Carole L. Cramer. But, she cautions, there are important differences as well. Human cells adorn some antibodies with special carbohydrate molecules. Plant cells can stick the wrong carbohydrates onto a human antibody. If that happens, says Douglas A. Russell, a molecular biologist at Agracetus, the maladjusted antibodies cannot stimulate the body into producing its own immune response, and they are rapidly filtered from the bloodstream. Until that discrepancy is solved, Russell says, Agracetus will focus on plantibodies that don't need the carbohydrates.

Could cloning resurrect the ancient mammoth? Is there a cell in a laboratory somewhere in the world that could bring the extinct dodo back to life? At this writing, the answer is no. But the fate of many endangered animals alive today might rely on some help from cloning.

To clone endangered animals, scientists use a technique where the chromosomes are removed from egg cells and are replaced by the embryonic skin cells of an endangered animal. When the egg is replaced in the womb and allowed to come to term, the result can be a clone of an endangered animal.

Cloning Noah's Ark

Robert P. Lanza, Betsy L. Dresser and Philip Damiani

In January 2001, a humble Iowa cow gave birth to the world's first cloned endangered species, a baby bull named Noah. Noah is a gaur: a member of a species of large oxlike animals that are now rare in their homelands of India, Indochina and southeast Asia. These one-ton bovines have been hunted for sport for generations. More recently the gaur's habitats of forests, bamboo jungles and grasslands have dwindled to the point that only roughly 36,000 are thought to remain in the wild. The World Conservation Union-IUCN Red Data Book lists the gaur as endangered, and trade in live gaur or gaur products—whether horns, hides or hooves-—is banned by the Convention on International Trade in Endangered Species (CITES).

While Noah died from a common bacterial infection, this birth was a new day in the conservation of his kind as well as in the preservation of many other endangered species. Perhaps most important, he was living, mooing proof that one animal can carry and give birth to the exact genetic duplicate, or clone, of an animal of a different species. And Noah was just

Gaur

the first creature up the ramp of the ark of endangered species that we and other scientists are currently attempting to clone: plans are under way to clone the African bongo antelope, the Sumatran tiger and that favorite of zoo lovers, the reluctant-to-reproduce giant panda. Cloning could also reincarnate some species that are already extinct—most immediately, perhaps, the bucardo mountain goat of Spain. The last bucardo—a female-died of a smashed skull when a tree fell on it, but Spanish scientists have preserved some of its cells.

Advances in cloning offer a way to preserve and propagate endangered species that reproduce poorly in zoos until their habitats can be restored and they can be reintroduced to the wild. Cloning's main power, however, is that it allows researchers to introduce new genes back into the gene pool of a species that has few remaining animals. Most zoos are not equipped to collect and cryopreserve semen; similarly, eggs are difficult to obtain and are damaged by freezing. But by cloning animals whose body cells have been preserved, scientists can keep the genes of that individual alive, maintaining (and in some instances increasing) the overall genetic diversity of endangered populations of that species.

Nevertheless, some conservation biologists have been slow to recognize the benefits of basic assisted reproduction strategies, such as in vitro fertilization, and have been hesitant to consider cloning. Although we agree that every effort should be made to preserve wild spaces for the incredible diversity of life that inhabits this planet, in some cases either the battle has already been lost or its outcome looks dire. Cloning technology is not a panacea, but it offers the opportunity to save some of the species that contribute to that diversity.

A clone still requires a mother, however, and very few conservationists would advocate rounding up wild female endangered animals for that purpose or subjecting a precious zoo resident of the same species to the rigors of assisted reproduction and surrogate motherhood. That means that to clone an endangered species, researchers such as ourselves must solve the problem of how to get cells from two different species to yield the clone of one.

A Gaur Is Born

It is a deceptively simple-looking process. A needle jabs through the protective layer, or zona pellucida, surrounding an egg that hours ago resided in a living ovary. In one deft movement, a research assistant uses it to suck out the egg's nucleus—which contains the majority of a cell's genetic material—leaving behind only a sac of gel called cytoplasm. Next he uses a second needle to inject another, whole cell under the egg's outer layer. With the flip of an electric switch, the cloning is complete: the electrical pulse fuses the introduced cell to the egg, and the early embryo begins to divide. In a few days, it will become a mass of cells large enough to implant into the uterus of a surrogate-mother animal previously treated with hormones. In a matter of months, that surrogate mother will give birth to a clone.

In practice, though, this technique—which scientists call nuclear transfer—is not so easy. To create Noah, Advanced Cell Technology (ACT) in Worcester, Mass., had to fuse skin cells taken from a male gaur with 692 enucleated cow eggs. Of those 692 cloned early embryos, only 81 grew in the laboratory into blastocysts, balls of 100 or so cells that are sufficiently developed to implant for gestation. We ended up inserting 42 blastocysts into 32 cows, but only eight became pregnant. We removed the fetuses from two of the pregnant cows for scientific analysis; four other animals experienced spontaneous abortions in the second or third month of the usual nine-month pregnancy; and the seventh cow had a very unexpected late-term spontaneous abortion in August 2000.

The statistics of the efficiency of cloning reflect the fact that the technology is still as much an art as it is a science—particularly when it involves transplanting an embryo into another species. Scientists, including those of us at ACT, have had the highest success rates cloning domestic cattle implanted into cows of the same species. But even in this instance we have had to work hard to produce just a few animals. For every 100 cow eggs we fuse with adult cattle cells, we can expect only between 15 and 20 to produce blastocysts. And only roughly 10 percent of those—one or two-yield live births.

The numbers reflect difficulties with the nuclear transfer process itself, which we are now working to understand. They are also a function of the vagaries of assisted reproduction technology.

Accordingly, we expect that the first few endangered species to be cloned will be those whose reproduction has already been well studied. Several zoos and conservation societies—including the Audubon Institute Center for Research of Endangered Species (AICRES) in New Orleans, which is led by one of us (Dresser)—have probed the reproductive biology of a range of endangered species, with some notable successes. Recently, Dresser and her colleagues reported the first transplantation of a previously frozen embryo of an endangered animal into another species that resulted in a live birth. In this case, an ordinary house cat gave birth to an African wildcat, a species that has declined in some areas.

So far, beyond the African wildcat and the gaur, we and others have accomplished interspecies embryo transfers in four additional cases: an Indian desert cat into a domestic cat; a bongo antelope into a more common African antelope called an eland; a mouflon sheep into a domestic sheep; and a rare red deer into a common white-tailed deer. All yielded live births. We hope that the studies of felines will pave the way for cloning the cheetah, of which only roughly 12,000 remain in southern Africa. The prolonged courtship behavior of cheetahs requires substantial territory, a possible explanation for why the animals have bred so poorly in zoos and yet another reason to fear their extinction as their habitat shrinks.

Panda-monium

One of the most exciting candidates for endangered-species cloning—the giant panda—has not yet been the subject of interspecies transfer experiments, but it has benefited from assisted reproduction technology. Following the well-publicized erotic fumblings of the National Zoo's ill-fated panda pair, the late Ling-Ling and Hsing-Hsing, the San Diego Zoo turned to artificial insemination to make proud parents of its Bai Yun and Shi Shi. Baby Hua Mei was born in August 1999.

Giant pandas are such emblems of endangered species that the World Wildlife Fund (WWF) uses one in its logo. According to a census that is now almost 20 years old, fewer than 1,000 pandas remain in their mountainous habitats of bamboo forest in southwest China. But some biologists think that the

THE NUCLEAR TRANSFER (CLONING) PROCESS

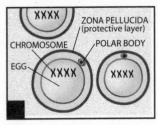 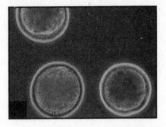

Recipient eggs are coaxed to mature in a culture dish. Each has a remnant egg cell called the polar body.

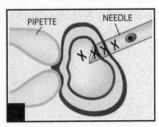 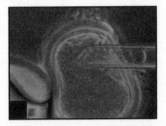

The polar bodies and chromosomes of each egg are drawn into a needle. A pipette holds the egg still.

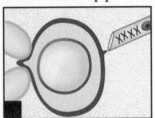 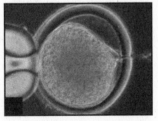

Once the chromosomes and polar body are removed, all that remains inside the zona pellucida is cytoplasm.

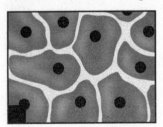 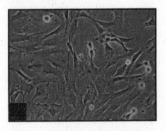

Skin cells called fibroblasts are isolated from the animal to be cloned and grown in culture dishes.

THE NUCLEAR TRANSFER (CLONING) PROCESS

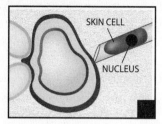 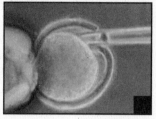

An entire skin cell is taken up into the needle, which is again punched through the zona pellucida.

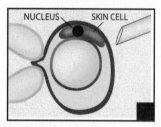 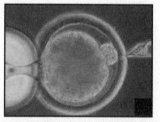

The skin cell is injected underneath the zona pellucida, where it remains separate from the egg cytoplasm.

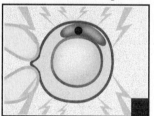 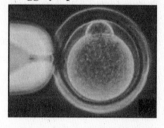

Each injected egg is exposed to an electric shock that fuses the skin cell with the egg cytoplasm.

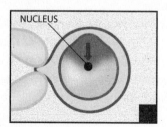 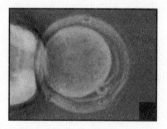

The skin cell's nucleus, with its genes, enters the egg cytoplasm. Within a few hours, the fused cell begins to divide.

population might have rebounded a bit in some areas. The WWF expects to complete a census of China's pandas in mid-2002 to produce a better estimate.

In the meantime, we at ACT are discussing plans with the government of China to clone a giant panda. Chinese scientists have already made strides toward the goal of panda cloning. In August 1999 Dayuan Chen of the institute and his co-workers published a paper in the English-language journal Science in China announcing that they had fused panda skeletal muscle, uterus and mammary gland cells with the eggs of a rabbit and then coaxed the cloned cells to develop into blastocysts in the laboratory.

A rabbit, of course, is too small to serve as a surrogate mother for a giant panda. Instead ACT and the Chinese plan to turn to American black bears. ACT has plans to obtain eggs from female black bears killed during hunting season in the northeastern U.S. Together with the Chinese, ACT scientists hope to use these eggs and frozen cells from the late Hsing-Hsing or Ling-Ling to generate cloned giant panda embryos that can be implanted into a female black bear now living in a zoo. A research group that includes veterinarians at Bear Country U.S.A. in Rapid City, S.D., has already demonstrated that black bears can give birth to transplanted embryos. They reported the successful birth of a black bear cub from an embryo transferred from one pregnant black bear to another last year in the journal Theriogenology.

AICRES scientists hope to take advantage of the success with bongo antelope that one of us (Dresser) had while at the Cincinnati Zoo. In 1984 Dresser and Charles Earle Pope and their colleagues announced the birth of a bongo after moving very early embryos from a pregnant female bongo to an eland surrogate mother.

Most of the mountain subspecies of bongo—a medium-size antelope with vertical white stripes—live in captivity. According to the World Conservation Union-IUCN, the mountain bongo is endangered, with only 50 or so remaining in a small region of Kenya. In contrast, the 1999 Bongo International Studbook lists nearly 550 mountain bongo living in zoos throughout the world. The lowland bongo subspecies is slightly better off: it is listed as "near threatened" and has a population of perhaps several thousand scattered throughout central and western Africa.

A coalition of conservation organizations in the U.S. and Kenya is now planning to send mountain bongo that have been bred in captivity to two sites in Kenya. And in a new approach to reintroducing a species, AICRES is working in Kenya to transfer frozen bongo embryos into eland surrogates. Cloning could support these efforts and possibly yield more bongo for reintroduction.

But what about animals that are already extinct? Chances are slim to nil that scientists will soon be able to clone dinosaurs, à la Jurassic Park, or woolly mammoths. The primary problem is the dearth of preserved tissue—and hence DNA. A group of researchers unearthed what they had hoped would be a well-preserved mammoth in 1999, but repeated freezing and thawing over the eons had poked holes in the creature's DNA, and molecular biologists have not yet found a feasible way of filling in such genetic gaps.

A similar difficulty has hobbled efforts by Australian scientists to clone a thylacine, or Tasmanian tiger, a wolflike marsupial that died out in the 1930s. Researchers at the Australian Museum in Sydney are attempting to clone cells from a thylacine pup that was preserved in alcohol in 1866, but the DNA is in such poor condition that they say they will have to reconstruct all of the animal's chromosomes.

The recently extinct bucardo may prove a more promising target for resurrection. ACT is arranging a collaboration with Alberto Fernández-Arias and José Folch of the Agricultural Research Service in Zaragoza, Spain. Fernández-Arias froze tissue from the last bucardo. He and Folch had tried for several

What About Rover and Fluffy?

The list of domesticated animals that scientists have been able to clone so far includes sheep, cattle, goats and laboratory mice. Compared with that menagerie, you'd think that cloning an ordinary dog or cat would be a snap. Unfortunately, this has not been the case. Both of our research groups have created cloned cat embryos and have implanted them into female cats, but as this article goes to press, neither of our teams has yet obtained a full-term pregnancy. Dogs have presented even more problems.

But we anticipate success soon. At Advanced Cell Technology (ACT), we have undertaken a research program that uses cloning technology to propagate pets as well as service animals such as seeing-eye dogs for the blind, hearing dogs for the deaf, search-and-rescue dogs, and animals used for social therapy. Together with Louisiana State University, the Audubon Institute has teamed up with a company called Lazaron BioTechnologies in Baton Rouge, La., to clone pet dogs and cats.

A surprising number of people are interested in cloning their favorite deceased pet in the hope of getting an animal with similar behavioral characteristics. A good deal of a cat or dog's demeanor is thought to be genetically determined. Although one can argue that there are already plenty of cats and dogs in the world that need homes, people still use traditional breeding methods to try to reproduce a particularly desirable animal. Cloning could offer a more efficient alternative. It could be particularly important in the case of service animals. Currently, for instance, male seeing-eye dogs are neutered at an early age so that they can concentrate better during their expensive and rigorous training. So, unfortunately, even if a dog turns out to be very good at his job, he can't be bred to produce more like him.

Our efforts to clone pets could also pay off for endangered species. We expect to be able to apply the information we obtain from cloning cats and dogs to preserving endangered felines and canines.

ACT and several other companies now offer pet cloning kits that veterinarians can use to preserve samples from a client's pet for possible future cloning. The kits contain materials for collecting a skin specimen and sending it back to a laboratory. Research assistants there use the tissue to establish a collection of pure, dividing cells called a cell line, which will be the source of donor cells for cloning. ACT extracts eggs for the cloning procedure from reproductive tracts taken from animals that have been spayed by veterinarians. We remove the ovaries and carefully puncture all visible follicles to release the eggs. Then we collect the eggs and place them in a specialized maturation medium that contains hormones, proteins and nutrients. Once fully matured, the eggs are ready for the nuclear transfer procedure.

So far our main focus has been the domestic cat, primarily because its reproductive physiology has been well studied, and embryo transfers of early- and late-stage embryos have resulted in the birth of live kittens. Both ACT and the Audubon Institute have been able to establish systems for prompting cat eggs to mature in the lab and have consistently produced cloned embryos that are being transferred to recipients.

But dogs are a different story. The dog's reproductive physiology is unique among mammalian species. Dogs ovulate an immature egg that has a very long maturation time. This means that we need a different maturation system from the one we have used in cats and that we have fewer eggs to work with in the end. So Fluffy will probably have a leg up on Rover when it comes to cloning.

years to preserve the mountain goat, which in the end was wiped out by poaching, habitat destruction and landslides. In 1999 they transferred embryos from a subspecies related to the bucardo to a domestic goat, yielding live kids.

But even if interspecies nuclear transfer succeeds for the bucardo, it will yield only a sorority of clones, because we have tissue from just one animal, a female. ACT plans to try to make a male by removing one copy of the X chromosome from one of the female bucardo's cells and using a tiny artificial cell called a microsome to add a Y chromosome from a closely related goat species. The technology has been used by other researchers to manipulate human chromosomes, but it has never before been used for cloning. A nonprofit organization called the Soma Foundation has been established to help fund such efforts.

Why Clone?

Cloning endangered species is controversial, but we assert that it has an important place in plans to manage species that are in danger of extinction. Some researchers have argued against it, maintaining that it would restrict an already dwindling amount of genetic diversity for those species. Not so. We advocate the establishment of a worldwide network of repositories to hold frozen tissue from all the individuals of an endangered species from which it is possible to collect samples. Those cells-like the sperm and eggs now being collected in "frozen zoos" by a variety of zoological parks-could serve as a genetic trust for reconstituting entire populations of a given species. Such an enterprise would be relatively inexpensive: a typical three-foot freezer can hold more than 2,000 samples and uses just a few dollars of electricity per year. Currently only AICRES and the San Diego Zoo's Center for Reproduction of Endangered Species maintain banks of frozen body cells that could be used for cloning.

Other critics claim that the practice could overshadow efforts to preserve habitat. We counter that while habitat preservation is the keystone of species conservation, some countries are too poor or too unstable to support sustainable conservation efforts. What is more, the continued growth of the human species will probably make it impossible to save enough habitat for some other species. Cloning by interspecies nuclear transfer offers the possibility of keeping the genetic stock of those species on hand without maintaining populations in captivity, which is a particularly costly enterprise in the case of large animals.

Another argument against cloning endangered species is that it might siphon donor money away from habitat maintenance. But not all potential donors are willing to support efforts to stem the tide of habitat destruction. We should recognize that some who would otherwise not donate to preserve endangered species at all might want to support cloning or other assisted reproduction technologies.

The time to act is now.

In 1997, a mere sheep touched off a whirlwind of media coverage and a storm of worldwide debate. And in many important ways, the storm is raging still.

Before Dolly, the hosts of other cloned animals had been generated from embryonic cells. Dolly was the first animal to be cloned from adult cells from an adult animal. The breakthrough constitutes an important step for the science of cloning, both from a technical perspective and from an ethical perspective. Technically at least, and after many tries (227 in the case of Dolly, to be exact), a new, genetically identical animal can be conjured from the cells of an existing animal. From an ethical perspective, the breakthrough puts grave new concerns on how the technology should be employed.

A Clone in Sheep's Clothing

Tim Beardsley

P hotographs of a rather ordinary-looking lamb named
Dolly made front pages around the world in 1997
because of her startling pedigree: Dolly, unlike any other
mammal that has ever lived, is an identical copy of another
adult and has no father. She is a clone, the creation of a group
of veterinary researchers. That work, performed by Ian Wilmut
and his colleagues at the Roslin Institute in Edinburgh, Scotland, has provided an important new research tool and has
shattered a belief widespread among biologists that cells from
adult mammals cannot be persuaded to regenerate a whole
animal. Although the Scottish researchers have made clear that
they would consider it unethical to adapt their technique to
clone humans (Wilmut is a member of a working group on the
ethics of genetic engineering), the demonstration has raised
the uncomfortable prospect that others might not be so
scrupulous. Cloning humans would mean that women could in
principle reproduce without any help from men [See "I,
Clone," page 89].

Wilmut and his co-workers accomplished their feat by transferring the nuclei from various types of sheep cells into unfertilized sheep eggs from which the natural nuclei had been removed by microsurgery. Once the transfer was complete, the recipient eggs contained a complete set of genes, just as they would if they had been fertilized by sperm. The eggs were then cultured for a period before being implanted into sheep that carried them to term, one of which culminated in a successful birth. The resulting lamb was, as expected, an exact genetic copy, or clone, of the sheep that provided the transferred nucleus, not of those that provided the egg.

Other researchers have previously cloned animals, including mammals, by transferring nuclei from embryonic cells into such enucleated eggs [see "Cloning Noah's Ark," page 23]. The interest in the new work, published in the February 27, 1997, *Nature*, is that some of the transferred nuclei that gave rise to lambs came not from embryonic cells but from the mammary gland of a mature, 6-year old ewe. Other workers have failed to produce viable offspring when they attempted equivalent experiments.

The key to success at the Roslin Institute seems to have been that Wilmut starved the mammary cells for five days before extracting their nuclei. This maneuver "froze" the cells in a quiescent phase of their division cycle and may have made their chromosomes more susceptible to being reprogrammed to initiate the growth of a new organism after the nuclei were transferred into an egg.

Wilmut's work is supported by a biotechnology company, PPL Therapeutics in Edinburgh, which plans to use the patented cloning technique to produce animals that will secrete valuable drugs in their milk [See "Transgenic Livestock as Drug Factories," page 57]. Other researchers engaged in similar work note that it is unclear how much practical benefit Wilmut's technique will yield in the short term: it is very labor-intensive and it required 277 nuclear transfers to produce the single, viable cloned lamb. At present, cloning from embryonic cells and even old-fashioned animal breeding are still more efficient ways of producing large numbers of genetically-altered animals, notes William H. Velander of Virginia Poly-technic Institute. Nor is it certain that the technique used to create Dolly can be applied to other species.

Even so, Wilmut's experiment provides a long-sought confirmation that adult cells do in fact contain workable versions of all the genes necessary to produce an entire organism. Moreover, the procedure will surely be refined and may become an important aid in all manner of biological and biomedical investigations. It might, for example, be used to mass-produce animals that mimic human diseases for research purposes [See "Cloning for Medicine," page 45]. The technique might in time also be used to improve livestock.

As for the possible use of cloning to produce copies of humans, most ethicists' initial reaction is that such an action would be unconscionable—although in the U.S., unlike in the U.K. and many other nations, it is not explicitly illegal. And opinions may change when confronted with real-world situations. Should grieving parents be denied the opportunity to produce an identical copy of their dying baby?

Shortly after the birth of Dolly, the first animal cloned from adult cells, another important development shook the science of cloning. After using ordinary adult cells to seed Dolly, genetically altered cells were used to produce genetically altered animals.

Commercial interests rather than pure research fueled this development. It suggests that in the not-so-distant future, identical copies of prize-winning bulls or goats designed to produce beneficial medical treatments could be in the offing.

Cloning Hits the Big Time

Tim Beardsley

When scientists at the Roslin Institute in Edinburgh reported in February 1997 that they had cloned an adult sheep, their work immediately captivated the public.

But some cautious researchers wondered whether the success might not be a flash in the pan. After all, the Scottish workers had to try 277 times before they succeeded in producing the clone, named Dolly. Unless the efficiency of the cloning process could be greatly improved, it seemed unlikely to become a common technique for producing improved strains of livestock. Moreover, Dolly's birth did not prove that cloning could be used to create animals from cells that had been genetically manipulated. Unless cloning could be combined with sophisticated genetic manipulation, the technique seemed unlikely to realize its full potential.

Now it is clear that those doubts are unfounded. While the debate focused on Dolly, a number of other corporate and academic laboratories were quietly pushing ahead with similar projects. At least two U.S. companies—ABS Global of De Forest, Wisc., and Advanced Cell Technology of Worcester, Mass.—

have successfully impregnated cows and pigs using cloned cells. In addition, recent work has demonstrated that cloning works perfectly well on cells that have been genetically altered.

Both corporate contenders claim that their cloning techniques are highly efficient. And neither is making any bones about its commercial intentions. In early August 1997, ABS Global simultaneously introduced a six month old bull named Gene and announced it had formed a new subsidiary, called Infigen, to "commercialize applications of cloning technologies in the cattle breeding, pharmaceutical, nutraceutical and xeno-transplantation fields."

Separately, Neal First of the University of Wisconsin at Madison has established, at least transiently, pregnancies in five different species using cloned adult cells. First says he has developed a "universal cloning system" based on cow egg cells that he has used to impregnate cows, sheep, rats, pigs and monkeys.

Researchers at Infigen use a variation on the Roslin approach, explains research director Michael D. Bishop. Rather than transferring just the nucleus, the ABS workers fuse a whole donor cell with an enucleated (nucleus removed) egg cell, a process that is helped along with a jolt of electricity. These donor cells are relatively unspecialized cells taken from fetuses.

When the resulting embryo has divided into about sixteen cells, it is broken up, or disaggregated, into its component cells. The resulting cells are themselves fused with other enucleated egg cells. These second-generation cells are then implanted into foster mothers to develop, which many of them do successfully. The calf "Gene" was cloned from fetal cells using this technique. By selectively making copies of genetically superior animals, Infigen's corporate parent hopes to boost its share of the lucrative market for bull semen. It might eventually begin selling cloned, genetically-altered animals, says Bishop.

Advanced Cell Technology, for its part, has initiated dozens of clone pregnancies in cows and some in pigs. For these clones, the donor cells were fibroblasts taken from fetuses. The genomes of these cells can be relatively easily and precisely manipulated through a technique known as targeted gene replacement. "Advanced Cell Technology has the ability to produce transgenic animals using fetal fibroblast nuclear transfer," claims Steve Parkinson, president and chief executive officer.

Parkinson contends that targeted gene replacement produces cells having specific genetic alterations far more effectively than the traditional technique for making transgenic animals, which entails injecting DNA into cell nuclei. He reports that Advanced Cell Technology plans to clone genetically altered animals whose neural tissue would be immunologically compatible with that of humans.

Cloning progress is not restricted to the U.S. Since the February 1997 breakthrough, PPL Therapeutics of Edinburgh, which collaborates with the Roslin Institute, has produced five lambs from fetal cells that were genetically modified to carry marker genes and genes for human proteins. Lambs produced from the genetically manipulated cells produce foreign proteins; such animals may be able to manufacture large quantities of medically valuable human proteins in their milk [see "Transgenic Livestock as Drug Factories, page 57]. The result "brings nearer the human benefits from nuclear transfer work," says Ron James, managing director of PPL. The same company is also working on cloned cattle in the U.S.

Dolly, then, was more than just an overnight sensation. Rather, cloning seems set to become a vital technology for agriculture and medicine. "I think the possibility is there that it might really move large-animal transgenic work forward much more rapidly," says Vernon Pursell of the U.S. Department of Agriculture. In other words, better forget the jokes and starting looking at the stock prices.

Ian Wilmut of the Roslin Institute in Edinburgh, Scotland spear-headed the efforts to create Dolly. In this article, he outlines the procedures he used in his landmark birth. He also extrapolates creative avenues for exploiting this new technology.

As one of the pioneers in the art of cloning animals from existing adults, Wilmut stress that caution and good moral sense are required if the field is to be treated as a valid and powerful branch of the biosciences.

Cloning for Medicine

Ian Wilmut

In the summer of 1995 the birth of two lambs at my institution, the Roslin Institute near Edinburgh in Midlothian, Scotland, heralded what many scientists believe will be a period of revolutionary opportunities in biology and medicine. Megan and Morag, both carried to term by a surrogate mother, were not produced from the union of a sperm and an egg. Rather their genetic material came from cultured cells originally derived from a nine-day-old embryo. That made Megan and Morag genetic copies, or clones, of the embryo.

Before the arrival of the lambs, researchers had already learned how to produce sheep, cattle and other animals by genetically copying cells painstakingly isolated from early-stage embryos. Our work promised to make cloning vastly more practical, because cultured cells are relatively easy to work with. Megan and Morag proved that even though such cells are partially specialized, or differentiated, they can be genetically reprogrammed to function like those in an early embryo. Most biologists had believed that this would be impossible.

We went on to clone animals from cultured cells taken from a 26-day-old fetus and from a mature ewe. The ewe's cells gave rise to Dolly, the first mammal to be cloned from an adult. Our announcement of Dolly's birth in February 1997 attracted enormous press interest, perhaps because Dolly drew attention to the theoretical possibility of cloning humans. This is an outcome I hope never comes to pass. But the ability to make clones from cultured cells derived from easily obtained tissue should bring numerous practical benefits in animal husbandry and medical science, as well as answer critical biological questions.

How to Clone

Cloning is based on nuclear transfer, the same technique scientists have used for some years to copy animals from embryonic cells. Nuclear transfer involves the use of two cells. The recipient cell is normally an unfertilized egg taken from an animal soon after ovulation. Such eggs are poised to begin developing once they are appropriately stimulated. The donor cell is the one to be copied. A researcher working under a high-power microscope holds the recipient egg cell by suction on the end of a fine pipette and uses an extremely fine micropipette to suck out the chromosomes, sausage-shaped bodies that incorporate the cell's DNA. (At this stage, chromosomes are not enclosed in a distinct nucleus.) Then, typically, the donor cell, complete with its nucleus, is fused with the recipient egg. Some fused cells start to develop like a normal embryo and produce offspring if implanted into the uterus of a surrogate mother.

In our experiments with cultured cells, we took special measures to make the donor and recipient cells compatible. In particular, we tried to coordinate the cycles of duplication of DNA and those of the production of messenger RNA, a molecule that is copied from DNA and guides the manufacture of proteins. We chose to use donor cells whose DNA was not being duplicated at the time of the transfer. To arrange this, we worked with cells that we forced to become quiescent by reducing the concentration of nutrients in their culture medium. In addition, we delivered pulses of electric current to the egg after the transfer, to encourage the cells to fuse and to mimic the stimulation normally provided by a sperm.

After the birth of Megan and Morag demonstrated that we could produce viable offspring from embryo-derived cultures, we filed for patents and started experiments to see whether offspring could be produced from more completely differentiated cultured cells. Working in collaboration with PPL Therapeutics, also near Edinburgh, we tested fetal fibroblasts (common cells found in connective tissue) and cells taken from the udder of a ewe that was three and a half months pregnant. We selected a pregnant adult because mammary cells grow vigorously at this stage of pregnancy, indicating that they might do well in culture. Moreover, they have stable chromosomes, suggesting that they retain all their genetic information. The successful cloning of Dolly from the mammary-derived culture and of other lambs from the cultured fibroblasts showed that the Roslin protocol was robust and repeatable.

All the cloned offspring in our experiments looked, as expected, like the breed of sheep that donated the originating nucleus, rather than like their surrogate mothers or the egg donors. Genetic tests prove beyond doubt that Dolly is indeed a clone of an adult. It is most likely that she was derived from a fully differentiated mammary cell, although it is impossible to be certain because the culture also contained some less differentiated cells found in small numbers in the mammary gland. Other laboratories have since used an essentially similar technique to create healthy clones of cattle and mice from cultured cells, including ones from nonpregnant animals.

Although cloning by nuclear transfer is repeatable, it has limitations. Some cloned cattle and sheep are unusually large, but this effect has also been seen when embryos are simply cultured before gestation. Perhaps more important, nuclear transfer is not yet efficient. John B. Gurdon, now at the University of Cambridge, found in nuclear-transfer experiments with frogs almost 30 years ago that the number of embryos surviving to become tadpoles was smaller when donor cells were taken from animals at a more advanced developmental stage. Our first results with mammals showed a similar pattern. All the cloning studies described so far show a consistent pattern of deaths during embryonic and fetal development, with laboratories reporting only 1 to 2 percent of embryos surviving to become live offspring. Sadly, even some clones that survive through birth die shortly afterward.

Clones with a Difference

The cause of these losses remains unknown, but it may reflect the complexity of the genetic reprogramming needed if a healthy offspring is to be born. If even one gene inappropriately expresses or fails to express a crucial protein at a sensitive point, the result might be fatal. Yet reprogramming might involve regulating thousands of genes in a process that could involve some randomness. Technical improvements, such as the use of different donor cells, might reduce the toll.

The ability to produce offspring from cultured cells opens up relatively easy ways to make genetically modified, or transgenic, animals. Such animals are important for research and can produce medically valuable human proteins.

The standard technique for making transgenic animals is painfully slow and inefficient. It entails microinjecting a genetic construct—a DNA sequence incorporating a desired gene—into a large number of fertilized eggs. A few of them take up the introduced DNA so that the resulting offspring express it. These animals are then bred to pass on the construct [see "Transgenic Livestock as Drug Factories," page 57].

In contrast, a simple chemical treatment can persuade cultured cells to take up a DNA construct. If these cells are then used as donors for nuclear transfer, the resulting cloned offspring will all carry the construct. The Roslin Institute and PPL Therapeutics have already used this approach to produce transgenic animals more efficiently than is possible with microinjection.

We have incorporated into sheep the gene for human factor IX, a blood-clotting protein used to treat hemophilia B. In this experiment we transferred an antibiotic-resistance gene to the donor cells along with the factor IX gene, so that by adding a toxic dose of the antibiotic neomycin to the culture, we could kill cells that had failed to take up the added DNA. Yet despite this genetic disruption, the proportion of embryos that developed to term after nuclear transfer was in line with our previous results.

The first transgenic sheep produced this way, Polly, was born in the summer of 1997. Polly and other transgenic clones secrete the human protein in their milk. These observations suggest that once techniques for the retrieval of egg cells in different species have been perfected, cloning will make it possible to introduce precise genetic changes into any mammal and to create multiple individuals bearing the alteration.

Cultures of mammary gland cells might have a particular advantage as donor material. Until recently, the only practical way to assess whether a DNA construct would cause a protein to be secreted in milk was to transfer it into female mice, then test their milk. It should be possible, however, to test mammary cells in culture directly. That will speed up the process of finding good constructs and cells that have incorporated them so as to give efficient secretion of the protein.

Cloning offers many other possibilities. One is the generation of genetically modified animal organs that are suitable for transplantation into humans. At present, thousands of patients die every year before a replacement heart, liver or kidney becomes available. A normal pig organ would be rapidly destroyed by a "hyperacute" immune reaction if transplanted into a human. This reaction is triggered by proteins on the pig cells that have been modified by an enzyme called alpha-galactosyl transferase. It stands to reason, then, that an organ from a pig that has been genetically altered so that it lacks this enzyme might be well tolerated if doctors gave the recipient drugs to suppress other, less extreme immune reactions.

Another promising area is the rapid production of large animals carrying genetic defects that mimic human illnesses, such as cystic fibrosis. Although mice have provided some information, mice and humans have very different genes for cystic fibrosis. Sheep are expected to be more valuable for research into this condition, because their lungs resemble those of humans. Moreover, because sheep live for years, scientists can evaluate their long-term responses to treatments.

Creating animals with genetic defects raises challenging ethical questions. But it seems clear that society does in the main support research on animals, provided that the illnesses being studied are serious ones and that efforts are made to avoid unnecessary suffering.

The power to make animals with a precisely engineered genetic constitution could also be employed more directly in cell-based therapies for important illnesses, including Parkinson's disease, diabetes and muscular dystrophy. None of these conditions currently has any fully effective treatment. In each, some pathological process damages specific cell populations, which are unable to repair or replace themselves. Several novel approaches are now being explored that would provide new cells—ones taken from the patient and cultured, donated by other humans or taken from animals.

To be useful, transferred cells must be incapable of transmitting new disease and must match the patient's physiological need closely. Any immune response they produce must be manageable. Cloned animals with precise genetic modifications that minimize the human immune response might constitute a plentiful supply of suitable cells. Animals might even produce cells with special properties, although any modifications would risk a stronger immune reaction.

Cloning could also be a way to produce herds of cattle that lack the prion protein gene. This gene makes cattle susceptible to infection with prions, agents that cause bovine spongiform encephalitis (BSE), or mad cow disease. Because many medicines contain gelatin or other products derived from cattle, health officials are concerned that prions from infected animals could infect patients. Cloning could create herds that, lacking the prion protein gene, would be a source of ingredients for certifiable prion-free medicines.

The technique might in addition curtail the transmission of genetic disease. Many scientists are now working on therapies that would supplement or replace defective genes in cells, but even successfully treated patients will still pass on defective genes to their offspring. If a couple was willing to produce an embryo that could be treated by advanced forms of gene therapy, nuclei from modified embryonic cells could be transferred to eggs to create children who would be entirely free of a given disease.

Some of the most ambitious medical projects now being considered envision the production of universal human donor cells. Scientists know how to isolate from very early mouse embryos undifferentiated stem cells, which can contribute to all the different tissues of the adult. Equivalent cells can be obtained for some other species, and humans are probably no exception. Scientists are learning how to differentiate stem cells in culture, so it may be possible to manufacture cells to repair or replace tissue damaged by illness.

Making Human Stem Cells

Stem cells matched to an individual patient could be made by creating an embryo by nuclear transfer just for that purpose, using one of the patient's cells as the donor and a human egg as the recipient. The embryo would be allowed to develop only to the stage needed to separate and culture stem cells from it. At that point, an embryo has only a few hundred cells, and they have not started to differentiate. In particular, the nervous system has not begun to develop, so the embryo has no means of feeling pain or sensing the environment. Embryo-derived cells might be used to treat a variety of serious diseases caused by damage to cells, perhaps including AIDS as well as Parkinson's, muscular dystrophy and diabetes.

Scenarios that involve growing human embryos for their cells are deeply disturbing to some people, because embryos have the potential to become people. The views of those who consider life sacred from conception should be respected, but I suggest a contrasting view. The embryo is a cluster of cells that does not become a sentient being until much later in development, so it is not yet a person. In the U.K., the Human Genetics Advisory Commission has initiated a major public consultation to assess attitudes toward this use of cloning.

Creating an embryo to treat a specific patient is likely to be an expensive proposition, so it might be more practical to establish permanent, stable human embryonic stem-cell lines from cloned embryos. Cells could then be differentiated as needed. Implanted cells derived this way would not be genetically perfect matches, but the immune reaction would probably be controllable. In the longer term, scientists might be able to develop methods for manufacturing genetically matched stem cells for a patient by "dedifferentiating" them directly, without having to utilize an embryo to do it [See "Mother Nature's Menders," page 69].

Several commentators and scientists have suggested that it

might in some cases be ethically acceptable to clone existing people. One scenario envisages generating a replacement for a dying relative. All such possibilities, however, raise the concern that the clone would be treated as less than a complete individual, because he or she would likely be subjected to limitations and expectations based on the family's knowledge of the genetic "twin." Those expectations might be false, because human personality is only partly determined by genes. The clone of an extrovert could have a quite different demeanor. Clones of athletes, movie stars, entrepreneurs or scientists might well choose different careers because of chance events in early life.

Some have also put forward the notion that couples in which one member is infertile might choose to make a copy of one or the other partner. But society ought to be concerned that a couple might not treat naturally a child who is a copy of just one of them. Because other methods are available for the treatment of all known types of infertility, conventional therapeutic avenues seem more appropriate. None of the suggested uses of cloning for making copies of existing people is ethically acceptable to my way of thinking, because they are not in the interests of the resulting child. It should go without saying that I strongly oppose allowing cloned human embryos to develop so that they can be tissue donors.

It nonetheless seems clear that cloning from cultured cells will offer important medical opportunities. Predictions about new technologies are often wrong: societal attitudes change; unexpected developments occur. Time will tell. But biomedical researchers probing the potential of cloning now have a full agenda.

Is *Quiescence* the Key to Cloning?

All the cells that we used as donors for our nuclear-transfer experiments were quiescent—that is, they were not making messenger *RNA*. Most cells spend much of their life cycle copying *DNA* sequences into messenger RNA, which guides the production of proteins. We chose to experiment with quiescent cells because they are easy to maintain for days in a uniform state. But Keith H. S. Campbell of our team recognized that they might be particularly suitable for cloning.

He conjectured that for a nuclear transfer to be successful, the natural production of RNA in the donor nucleus must first be inhibited. The reason is that cells in a very early stage embryo are controlled by proteins and RNA made in the precursor of the parent egg cell. Only about three days after fertilization does the embryo start making its own RNA. Because an egg cell's own chromosomes would normally not be making RNA, nuclei from quiescent cells may have a better chance of developing after transfer.

A related possibility is that the chromosomes in quiescent nuclei may be in an especially favorable physical state. We think regulatory molecules in the recipient egg act on the transferred nucleus to reprogram it. Although we do not know what these molecules are, the chromosomes of a quiescent cell may be more accessible to them.

In many ways, genetic manipulation and cloning go hand in hand. Once the manipulation of a cell's genetic makeup is complete, it can be inserted into a donor egg and returned to the womb. If all goes well the result is an animal that expresses the genetic changes engineered by the scientist.

When human genes are blended with animal genes and then cloned, the resulting creature is known as a transgenic animal. At present, work in this arena aims to harvest a host of beneficial medical compounds from the milk and other products of transgenic animals.

Transgenic Livestock as Drug Factories

William H. Velander, Henryk Lubon and William N. Drohan

Exactly one year after her own birth, Genie, our experimental sow, was serenely nursing seven healthy piglets, her milk providing the many nutrients these offspring needed to survive and grow. But unlike other pigs, Genie's milk also contained a substance that some seriously ill people desperately need: human protein C. Traditional methods of obtaining such blood proteins for patients involve processing large quantities of donated human blood or culturing vast numbers of cells in giant stainless-steel reactor vessels. Yet Genie was producing copious amounts of protein C without visible assistance. She was the world's first pig to produce a human protein in her milk.

Genie's ability to manufacture a therapeutic drug in this way was the outcome of a research project conceived almost a decade ago. In collaboration with scientists from the American Red Cross who specialized in providing such blood proteins, we began to consider the possibility of changing the composition of an animal's milk to include some of these critically needed substances. In theory, this approach could generate any required quantity of the various therapeutic blood proteins that are regularly in short supply.

Demand for such drugs comes from many quarters. For instance, hemophiliacs may lack any of several different clotting agents, particularly blood proteins called Factor VIII and Factor IX. Certain people with an inborn deficiency require extra protein C (which acts to control clotting) to supplement their body's meager stores, and patients undergoing joint replacement surgery can benefit from this protein as well. Another important example of the need for therapeutic blood proteins involves people suffering strokes or heart attacks: these cases often demand quick treatment with a protein called tissue plasminogen activator, a substance that can dissolve blood clots. And some people suffering from a debilitating form of emphysema can breathe more easily with infusions of a protein called alpha-1-antitrypsin.

All these proteins are present in donated blood only in tiny amounts, and hence they are currently so difficult to produce that their expense precludes or severely limits their use as drugs. For example, treatment with purified Factor VIII (restricted to those times when someone with hemophilia is actually bleeding) typically costs the patient tens of thousands of dollars every year. The cost of continuous replacement of this blood protein for the same period—a desirable but rarely available option—would exceed $100,000.

Such enormous sums reflect the many problems involved in extracting these proteins from donated blood or establishing specialized production facilities using cultured cells—an enterprise that can require an investment of $25 million or more to supply even modest amounts of a single type of protein. Developing "transgenic" animals such as Genie (that is, creatures that carry genes from other species) demands only a small fraction of such costs. Yet the new breeds simplify procedures enormously and can produce vast quantities of human blood protein. Replacing conventional bioreactors with transgenic livestock thus offers immense economic benefits.

Creating blood proteins in this fashion also stands to better the other current practice—purifying them from donated blood—because it would circumvent the risk of contamination with infectious agents. Although blood proteins derived from pooled blood plasma are considered relatively safe now that donors are carefully screened and virus inactivation treatments are routinely applied, the threat from some pathogens always looms. For example, the fear of inadvertently spreading HIV (the AIDS-causing agent) and the hepatitis C virus is spurring researchers to seek substitutes for drugs now derived from human blood. Similarly, recent concerns about Creutzfeldt-Jakob disease (a degenerative disease of the nervous system) has caused some blood products to be withdrawn from the U.S. and Europe. Creating human blood proteins with transgenic livestock that are known to be free of such diseases would deftly sidestep these difficulties.

The many gains that would result from the use of transgenic animals as bioreactors gave us ample reason to pursue our vision of tidy stalls occupied by healthy livestock carrying a few key human genes. But at the outset of our work, we had many worries about the technical hurdles facing us in breeding such transgenic animals and garnering usable quantities of protein from their milk. Fortunately, we were able to progress rapidly, benefiting from a body of trailblazing research that had already been done.

Prior Mousing Around

As early as 1980, Jon W. Gordon and his colleagues at Yale University had determined that a fertilized mouse embryo could incorporate foreign genetic material (DNA) into its chromosomes—the cellular storehouses of genetic material. Shortly afterward, Thomas E. Wagner and his associates at the University of Ohio demonstrated that a gene (a segment of

DNA that codes for a particular protein) taken from a rabbit could function in a mouse. Using a finely drawn glass tube of microscopic dimensions, these researchers devised a way to inject a specific fragment of rabbit DNA into a single-cell mouse embryo. Amazingly, that DNA would often become integrated into the mouse's chromosomes, perhaps because it was recognized by the cell as a broken bit of DNA that needed to be repaired.

These researchers then implanted the injected embryos in a surrogate mother mouse and found that some of the mice born to her contained the rabbit gene in all their tissues. These transgenic mice in turn passed the foreign gene on to their offspring in the normal manner, following Mendel's laws of inheritance. The added gene functioned normally in its new host, and these mice made rabbit hemoglobin in their blood.

Another milestone on the road to transgenic animal bioreactors was passed in 1987. Along with their respective colleagues, both Lothar Hennighausen of the National Institute for Kidney and Digestive Diseases and A. John Clark of the Institute of Animal Physiology and Genetics at the Edinburgh Research Station in Scotland established means for activating foreign genes in the mammary glands of mice. Foreign protein molecules created in this way were then secreted directly into a transgenic mouse's milk, where they could be easily collected. These researchers accomplished this feat by combining the foreign gene of interest with a short segment of DNA that normally serves to activate a gene for a mouse milk protein.

Whereas Hennighausen's mice produced the desired human protein (in that case, tissue plasminogen activator) at disappointingly low concentrations, Clark's mice produced 23 grams of a sheep milk protein (known as betalactoglobulin) in each liter of milk—approximately matching a mouse's own major milk proteins in abundance. But betalactoglobulin was not a human protein in short supply, nor were these tiny mice the proper vehicle to provide useful quantities of milk. So Clark and his colleagues went to work injecting sheep embryos with DNA that contained a medically important human gene. They used the gene that codes for a blood-clotting factor (Factor IX), along with a segment of sheep DNA that normally switches on the production of betalactoglobulin in the mammary gland. Two years later Clark's transgenic sheep secreted Factor IX in their milk—but at barely detectable levels. It was at that juncture that we began our attempts to realize the potential of such pioneering work. But we decided to take a gamble and try a novel strategy.

A Pig in a Poke

Whereas other research groups had picked sheep, goats or cows as suitable dairy animals for producing human proteins, we chose to work with pigs instead. Swine offer the advantages of short gestation periods (four months), short generational times (12 months) and large litter sizes (typically 10 to 12 piglets). Thus, producing transgenic pigs is relatively quick compared with transforming other types of livestock. And despite their lack of recognition as dairy animals, pigs do produce quite a lot of milk: a lactating sow generates about 300 liters in a year. The real question for us was whether this unconventional choice of transgenic animal could in fact be made to produce appreciable levels of human protein in its milk.

Toward that end, we decided to use a DNA segment made up of a human gene and the so-called promoter for a major mouse milk protein (called whey acidic protein) that had been characterized by Hennighausen and his colleagues. By injecting this DNA combination into mouse embryos, those researchers were able to augment a mouse's chromosomes so that the creature would produce the desired human protein in its milk. To take advantage of this approach, we, too, fashioned a fragment of DNA that contained the human gene for the target protein (in our case, protein C) and the mouse promoter for whey acidic protein. But we injected this DNA into a set of pig embryos.

By implanting these fertilized cells in a surrogate mother pig, we could identify—after four months of nervous waiting—a newborn female piglet that carried the foreign DNA in all its cells. But even with this accomplishment, we had to remain patient for another year as our transgenic piglet, Genie, matured. Only then could we find out whether she would indeed produce the human protein in her milk. To our delight, Genie's milk contained protein C. Although the human protein was not as abundant as some of the pig's own milk proteins, it was nonetheless present in substantial amounts, with about one gram of protein C in each liter of milk—200 times the concentration at which this protein is found in normal human blood plasma. But we were also anxious to find out if this pig-made human protein would be biologically active.

We were concerned because the details of protein synthesis inside cells remain somewhat mysterious. The workings of the cellular machinery for reading the genetic code and translating that information into a sequence of amino acids—the building blocks for protein molecules—are, for the most part, well understood by biologists. But there are some subtle manipulations that need to be done by cells after the amino acids are joined together. These so-called post-translational modifications give a newly constructed protein molecule the final shape and chemical composition it needs to function properly. Post-translational modifications require complex cellular operations to cut off parts of the protein and to paste various chemical groups onto the molecule as it is assembled. Would the cells of Genie's mammary tissue be able to carry out those modifications well enough to make a working version of the human blood protein?

To determine the answer, we had to tackle the new problem of isolating a human blood protein from pig milk. First we removed the milk fat by centrifugation. Then we purified the remaining whey using a procedure that would extract only the biologically active part of the human protein. To our amazement, this component amounted to about one third of the total complement of protein C present. Never before had functional protein C been produced and harvested at such high levels from a transgenic animal—or from a conventional bioreactor. Genie had passed a major test, providing the first practical demonstration that a complex human protein could be produced in the milk of livestock.

Next Year's Model?

We devoted several years to studying Genie and many of her extant offspring and then began to focus our efforts on increasing the concentration of active human protein in the milk. Our intent was to overcome the limitations of mammary tissue in making the needed post-translational modifications. In principle, breaking through those final barriers could triple the output of useful protein molecules produced.

With some painstaking research into the problem, we discovered that most of the protein C remained in an immature, inactive form because there were insufficient amounts of a key processing enzyme named furin—itself a complex protein—within these cells. Hence, we immediately asked ourselves whether we could improve the situation by introducing another foreign gene, one that would allow more of the needed processing enzyme to be made.

To test this possibility quickly, we switched our efforts temporarily from pig to mouse, the fast-breeding main-stay of most transgenic mammal experiments. In 1995 we succeeded in engineering a line of transgenic mice that contained two human genes—one for protein C and one for furin. We arranged for both of these transgenes to switch on in the mammary gland by attaching them to the DNA promoter we had previously incorporated in Genie.

After months of tedious effort in the lab, we were ecstatic to find that these mice were able to secrete the mature form of protein C in their milk. We have thus started development of a new and improved transgenic pig that contains human genes for both protein C and furin. We expect soon to see a pig that produces three times more active protein C than Genie did, and we anticipate that other researchers working with transgenic livestock will also be able to fashion genetic modifications that cause the manufacture of processing enzymes along with the target protein.

Chimerical Visions

The notion of obtaining essentially unlimited quantities of scarce human blood proteins at reasonable cost would have seemed pure fantasy just a short time ago. For more than two decades, molecular biologists and biochemical engineers have labored to overcome the problems of producing even modest amounts of human proteins from large-scale cell culture facilities. Yet making biological pharmaceuticals in huge stainless-steel vats of genetically engineered cells seemed destined to remain an awkward and expensive undertaking.

Such bioreactors are enormously costly to construct, and they prove in operation to be extremely sensitive to small changes in the temperature and composition of the broth in which the cells are grown. In contrast, transgenic livestock bioreactors can be created merely by breeding more animals. Transgenic livestock need only routine attention to control their living conditions and nutrient supply, and yet they can easily produce the desired proteins at much higher concentrations than their metallic counterparts.

Although some risk exists that pathogens could be transmitted from livestock to humans, formal procedures are available to establish pedigreed animals that are free of known diseases. Indeed, such specific-pathogen-free herds are a well-established part of the agriculture industry. In addition, decades of the clinical use of pigs to produce insulin for diabetics give us confidence that swine can readily serve as bio-reactors for therapeutic human proteins without presenting undue hazard.

Still, like all new medicines, the human proteins produced in this way need to be carefully tested for safety and effectiveness before the government approves them for widespread use. The first example to be so examined (an anticlotting protein called antithrombin III, manufactured by Genzyme Transgenics Corporation using transgenic goats) begin clinical trials.

It is possible that the subtle differences between human and animal cells in the way post-translational modifications are carried out may affect how such proteins function in people. For example, certain modifications cause proteins to be cleared from the blood quickly by the liver, and so we suspect that some of the differences between the animal and human forms of these proteins could actually constitute improvements in the way these substances function as long-lived therapeutic drugs.

It is tempting to view the development of transgenic livestock bioreactors purely as a triumph of technology. But the history of this science also highlights the limits of what people can do with sophisticated machines. The mammary gland is optimized to maintain a high density of cells, to deliver to them an ample supply of nutrients and to channel the valuable proteins produced into an easily harvested form. Mammary tissue proves far superior to any cell-culture apparatus ever engineered for these tasks. Despite all their efforts to improve industrial cell-culture facilities, it turns out that a generation of biochemical engineers were unable to match the abilities of a tool for making proteins that nature had already honed.

What's Good for Genie . . .

The advent of transgenic techniques for manipulating live-stock also raised legitimate concerns about the health and welfare of the animals altered in this rather unorthodox way. After all, engineered "transgenes" of the kind we implanted in pig embryos can ultimately become part of each and every cell of the mature animals. What if an introduced gene turns on inappropriately and produces the foreign protein in a way that dam-ages the surrounding tissue?

Such worries made it critically important that we design our genetic manipulations so that the foreign gene would be driven into action only in the mammary gland—that is, within tissues that have a natural ability to produce and export protein without harming themselves or their host. We could expect to achieve such targeted control of protein production in our transgenic pigs because we used a promoter from a milk gene—a genetic switch of a type that is present in all mammals. Yet we recognized that even such well-behaved genes can show some promiscuous activity.

The genes we introduced into pigs, for example, also produce small amounts of their foreign proteins in the animals' salivary glands. These tissues are, in fact, quite similar in composition to mammary tissue. So we fully expected this incidental production, and we are quite sure that this minor side effect does not harm the pigs in any way. The lack of detrimental side effects is crucial—for the animals involved and also for the success of this pioneering method. One of the primary reasons for developing transgenic livestock to supply human proteins is to limit the possibility of transmitting diseases to the recipients of these drugs. Using anything but the healthiest livestock to produce these substances could increase the animals' susceptibility to disease as well as the possibility that they might accidentally pass on some unknown pathogen. Genetically engineering weakened livestock would thus, in the end, only prove self-defeating in the quest to produce safe and plentiful medicines.

Perhaps one of the greatest hopes of researchers studying the mechanics of cloning is not the creation of genetically superior animals or the development of animals that produce human factors for medical treatments, but is the cloning of key cells in the human body known as stem cells.

Stem cells are undifferentiated cells generated by the human body that can become virtually any cell in the body. Thus, the normal wear and tear on the body is repaired smoothly, seamlessly, throughout our lives.

In the event of extraordinary damage or of disease, the production of stem cells could be boosted by cloning them. But, right now, the most flexible stem cells with the broadest range of applications are harvested from aborted or miscarried human embryos. Thus, at this early stage of investigation, a strong polarization of opinion has generated enormous controversy as to whether or not research into the medical application of stem cells should continue. As the following article explains, technical, legal, and moral issues present a formidable barrier.

had cloned an adult sheep
diately captivated the pub
some cautious researchers
her the success might not
the pan. After all, the Scot
ers had to try 277 times be
eeded in producing the clor
y. Unless the efficiency of
ing process could be greatl

Mother Nature's Menders

Mike May

I n the 1970s the television program *The Six Million Dollar Man* opened each week by showing a terrible accident that turned astronaut Steve Austin into "a man barely alive." Then we heard: "Gentlemen, we can rebuild him. We have the technology." The idea intrigued us but seemed centuries away. It's not. An explosion of work surrounding stem cells, which can differentiate into many other cell types, raises hope for medical repairs beyond our imagination—mending a damaged heart, fixing a failing liver, improving a forgetful brain and, most exciting, significantly extending life. Instead of using bionic parts, like the ones that made Steve Austin stronger and faster, this technology could provide us with longer and healthier lives by enabling us to control our natural repair mechanisms.

This emerging field takes advantage of a cell that may emerge from the moment of conception. When a sperm cell works its way into an egg during fertilization, some scientists consider the result to be a stem cell. Other researchers consider stem cells to appear after several cell divisions that turn a fertilized egg into a hollow sphere of cells called a blastocyst. That sphere includes a region called the inner cell mass, consisting of a group of stem cells. Wherever stem cells first arise, they can branch out in many directions.

Many of us imagine that a human body builds up most of its cells and tissues early in life, and then everything begins to fall apart, cell by cell. New findings prove otherwise. Stem cells busily work away throughout our lives, acting like an army of housekeepers, cleaning up a little mess here and repairing some damage there. In some cases, a group of these cells work together to perform gargantuan tasks. For example, the stem cells located in bone marrow must replace more than one billion red blood cells every day. Such rebuilding might be going on constantly all over the body. Stem cells also seem to make new cells continuously for bone, liver, heart, muscle and even the brain, where scientists long thought that we were incapable of generating new cells.

Bodily Tune-ups

Stem cells serve as a natural defense against aging. As things wear out, these cells can repair some damage. As we get older, though, the failures in our bodies apparently overrun the stem cells. Consequently, we decline—getting slower, weaker, more forgetful. Nevertheless, many scientists believe that they could slow these processes with a stem cell tune-up. Moreover, a regular dose of jazzed-up stem cells might fight off degeneration and keep us living a longer and healthier life.

The inherent qualities of stem cells have drawn tremendous attention to them. To be sure, some scientists take the Six Million Dollar Man approach and try to fabricate new parts from exotic metals and space-age polymers. You can already get an artificial hip joint, an implantable device to help with hearing loss, and replacement valves for your heart. Some groups are even pursuing an electronic retina. But why rely on so many different parts—essentially a new fix for every problem—when you could use stem cells instead? Stem cells might be a cure-all of sorts, basically one-stop shopping for repairing anything that ails you.

Despite the recent interest in stem cells, they are not entirely new in medical therapies. Physicians have been extending human lives for years by including stem cells in some treatments. For example, some forms of cancer, such as childhood leukemia, require such a devastating dose of chemotherapy that it destroys a patient's bone marrow. A bone marrow transplant can restore a patient's blood-making capability, presumably because it provides a new supply of blood-making stem cells. When physicians started using bone marrow transplants, though, no one had seen a human stem cell. They just assumed that such cells existed.

In late 1998 all that changed. Two sets of researchers in the U.S.—John Gearhart's group at Johns Hopkins University and James Thomson's team at the University of Wisconsin-Madison—isolated human stem cells. These results shook up science and society, raising hope for therapeutic uses of stem cells as well as a range of ethical questions.

After the first reports, investigators launched a parade of promising animal experiments. Evan Snyder of Harvard Medical School has shown that neural stem cells seek out damaged areas of a mouse's cortex—the highest centers of the brain—and make new neurons there. He has very preliminary evidence that neural stem cells can do this in primates, too. "We're starting to move our way up the evolutionary ladder," Snyder says, "suggesting that this really may be a kind of intervention or kind of application that we could use." He also mentions evidence that neural stem cells could generate new neurons in other areas of the brain and even in the spinal cord. If human neural stem cells can go to damaged areas in the nervous system and create neurons there, such a technique might fend off Parkinson's disease, amyotrophic lateral sclerosis (better known as Lou Gehrig's disease) or old-age dementia.

Tissue Flipping

These findings seem to be cropping up in one organ after another. For instance, Bryon Petersen of the University of Florida says his work in rats showed that a cell that originated in the bone marrow could travel to the liver, incorporate into that organ and become a functioning liver cell. Presumably, that bone marrow cell was a blood-making stem cell. As Ronald McKay of the National Institute of Neurological Disorders and Strokes explains, "One really exciting thing that's going on in the field at the moment is, in fact, we're sort of discovering that the stem cells that have been defined in different tissues are actually capable of flipping from one tissue to another." McKay notes that researchers are not absolutely sure that the flipping really goes on in stem cells but adds that "there are cells that are capable of giving rise to the cells of another tissue: brain into blood, brain into muscle, pancreas into liver, muscle into blood."

Still, scientists must answer a crucial question: Do the new cells really work? In most cases, it's hard to tell. Just because a stem cell ends up in the brain and turns into what looks exactly like a neuron doesn't mean that it works properly. Still, McKay and his colleagues did show at least one case in which new neurons did work. First, they caused a Parkinson's-like disease in rats by killing neurons that communicate through a neurotransmitter called dopamine. Then they obtained neural stem cells from rat embryos and injected the cells into the Parkinsonian adult rats. In less than three months the normal movement in most of the treated rats improved by about 75 percent.

Over this incredibly promising work looms a controversy that threatens some stem cell research. It all revolves around one word: embryo. In essence, scientists talk about two general classes of stem cells, ones that come from embryos and ones from adults. Some people would never condone using embryos in any way because of ethical beliefs. If you can get stem cells from adults, though, surely this entire problem can be resolved by forgoing the use of embryonic stem cells. But, as Thomson explains, "the embryonic stem cells have the potential to form anything. It's not clear what the developmental potential is of some of these other stem cells." In other words, an embryonic stem cell can do it all—make any cell needed—and adult stem cells might be limited to making a few kinds of cells. Furthermore, adult stem cells could be partially worn out, so that they would not offer the full rejuvenating benefits of embryonic ones.

Despite all the potential benefits of using embryonic stem cells, working with them remains off-limits for researchers receiving federal funding for their studies, as all powerful laboratories do. A ban put in place by the U.S. Congress on the use of federal money means that research is confined to the narrow universe of just a few private biotechnology companies. Progress in the field is slower than it might be without the prohibition. But the funding environment may change.

In November 1998 President Bill Clinton asked the National Bioethics Advisory Commission to investigate the medical and ethical issues behind embryonic stem cells. Its report concluded: "[T]he Commission believes that federal funding for the use and derivation of [embryonic stem] cells should be limited to two sources of such material: cadaveric fetal tissue [from naturally aborted fetuses] and embryos remaining after infertility treatments." The report thus encouraged federal funding for certain approaches to stem cell research. Then, in December 1999, the National Institutes of Health, the primary source of U.S. biomedical funding, published a draft for guidelines on stem cell research, which went out for public comment. These documents suggest that the outlook for at least limited federal support has become less bleak.

But the National Bioethics Advisory Commission did not endorse an approach called nuclear transfer, which Michael West of Advanced Cell Technology champions. In his technique, researchers remove the nucleus from a cow's egg, implant a human cell—say, a skin cell—inside it and allow it to grow embryonic stem cells. With this system, West and his colleagues might be able to use skin cells—obtained by merely scraping a toothpick across the inside of your cheek—to make embryonic stem cells just for you. That could be important because your immune system might fight off stem cells from anyone else, seeing them as foreign invaders, like a virus. In addition, cow's eggs come cheaply and in large numbers. Still, combining human and cow cells started more than a little disgruntled mooing, because some people see it as a dangerous mixing of species. West defends his approach, saying, "We take the [cow] egg and remove its DNA, so there's no more mixing of species than there is when you drink cow's milk."

In any case, you need West's approach only if your body really is likely to reject foreign stem cells. West says your immune system would search out foreign cells with the efficiency of a hawk hunting a mouse. Thomson, one of the first to isolate human embryonic stem cells, agrees that the body would reject stem cells as it does some organ transplants: "Absolutely. Once they differentiate, they'll become adult cells like any other cell in the body," which would cause them to be rejected. But Thomson's opinion is not universal. One company, Osiris Therapeutics, has found through its studies that foreign stem cells are not cast out by the immune system. "It really doesn't seem to be the case. We don't quite know why that is," says Osiris scientist Mark Pittenger.

Luckily, investigators do agree on some topics. For example, most everyone thinks they could grow these cells in culture and keep them alive essentially forever. About his human embryonic cells, for instance, Thomson says, "We've kept them growing for well over a year. By any measure that we have, they appear to be immortal." And once researchers know how to culture whatever kind of cells they have, they can make incredibly large numbers of them. For example, a single human skin cell can spawn 170 trillion trillion trillion cells. Moreover, farming these cells in culture could reduce the concerns about using embryonic tissue. "One of the things I think people don't like about this is the idea of constantly going back to human embryos and doing the stuff over and over and over again," McKay says. "But technically, we can grow the cells, we can really grow them. I think this is going to be very efficient, so you needn't be concerned that this is going to be a big [embryo] harvesting industry. It's not going to be like that." The recently created WiCell Research Institute—with Thomson as its scientific director—plans to grow and sell human embryonic stem cells for research.

Although physicians already rely somewhat on stem cells—at least for bone marrow transplants—many more clinical applications might lie just over the horizon. Stem cells might be used to repopulate or replace cells devastated by disease. It might even be possible to take a stem cell, nudge it chemically toward making the kind of tissue desired and then control its environment in a way that causes it to build an entire organ. The organ could then be used in someone who needs a transplant, the pinnacle of so-called tissue engineering.

When could some of these stem cell techniques be available? "I think we're going to be moving into clinical trials with human neural stem cells of some type for some disease within two years," Snyder says. That means that stem cell-boosting treatments could be available in five to 10 years. Making entire organs from scratch, however, lies much further in the future. In theory, physicians could get so good at fixing organs with stem cell treatments that such organ fabrication might never be needed.

Just Hit "Play"

Some companies are already counting on a market in stem cell medicine. For instance, Douglas Armstrong of Aastrom Biosciences describes a machine developed by his company that is primed with a sample of bone marrow or even blood from an umbilical cord, both of which contain stem cells. According to Armstrong, "The equipment operates much like a VCR with a videocassette. The user takes the cassette, pops it into the machine and the machine takes over. Twelve days later the cassette comes back out, goes on another machine and transfers the cell product to a blood bag that's ready for therapy."

The resulting blood bag would contain stem cells as well as so-called progenitor cells, which are products of stem cells that are primed to make specific tissues. A physician could simply inject stem and progenitor cells into the bloodstream, and many of them would home in on locations where they were needed.

Armstrong adds, "It's practical to think we may be entering a future period where all of us put aside a small amount of bone marrow or even our umbilical cord blood when we are born, and then samples of that are grown out into populations and we get infusions of those cells later in life that might, indeed, help us live much longer, healthier lives."

Many hurdles lie between ongoing research and turning stem cell techniques into therapies for humans. "These therapies are brand-new," Thomson says. "There are no precedents for them." Consequently, a researcher can't simply see what stem cell treatments do in rats and mice and then try the same thing in humans. In a hypothetical example, Thomson speculates that a newly discovered technique that cures diabetes in mice might not help human diabetics but instead leave them with a worse disease—pancreatic cancer, for example. In other words, a treatment for a serious but survivable disease could give patients a certain death sentence.

"So you want to make really sure what you're doing isn't worse than the disease you're trying to cure and that there's a lot of safety involved," Thomson continues. "Because the therapies are so new, going straight into humans would be a problem." Much more research on primates and then extensive clinical testing must be completed before new stem cell techniques become available as a routine form of treatment.

Scientists do know that stem cells promise entirely new views of how the human body works. "For me, the absolute true potential of these is more in how it's going to give us a clue to understand the human body," Thomson says. "So even if [stem cells] were never to be used for transplantation purposes, they give you this brand-new scientific model to study. If you're interested in heart disease, you can study populations of human heart cells in tissue culture for the first time on a regular basis.

"I think the transplantation stuff will be important," he goes on, "but someday we'll understand enough about the human body that these transplantation therapies won't be necessary, because it will be possible to cause specific cells to regenerate themselves in ways they don't naturally do, because we will understand how that development normally occurs."

We might never see science rebuild a man with Steve Austin-like techniques. Instead researchers may rebuild us by tweaking systems that our bodies possessed all along—stem cells, the ultimate medical weapons. Now we must wait to see if science and society can agree on ways to use these seemingly magical wonders of biology.

The rejection of donor organs has bedeviled the advancement of medical science for years. Now that researchers have a better handle on the causes of rejection, one of the more promising avenues for a solution lies in cloning.

By pinpointing the causes of hyperreaction to foreign tissue, genetic engineers can selectively remove or modify them. The resulting cells could then be placed in a donor egg and used to create an animal whose organs would dovetail with the human body.

Xenotransplantation

Robert P. Lanza,
David K.C. Cooper and
William L. Chick

Early morning, sometime in the near future. A team of surgeons removes the heart, lungs, liver, kidneys and pancreas from a donor, whereupon a medical technician packs these organs in ice and rushes them to a nearby airport. A few hours later the heart and liver land in one city, the two kidneys in another, and the lungs and pancreas arrive in a third. Speedily conveyed to hospitals in each city, these organs are transplanted into patients who are desperately ill. The replacements function well, and six people receive a new lease on life. Back at the donor center, surgeons repeat the procedure several times, and additional transplants take place at a score of facilities distributed around the country. In all, surgical teams scattered throughout the U.S. conduct more than 100 transplant operations on this day alone.

How could so many organ donors have possibly been found? Easily—by obtaining organs not from human cadavers but from pigs. Although such a medical miracle is not yet possible, we and other researchers are taking definite steps toward it. Our efforts are driven by the knowledge that the supply of human organs will always be insufficient to satisfy demand. Within just the U.S., thousands of patients await transplants of the heart, liver, kidney, lung and pancreas, and millions struggle with diseases that may one day be curable with other kinds of donations. Notably, hemophilia, diabetes and even Alzheimer's and Parkinson's diseases might well be treated using transplanted cells. So the pressure to devise ways to transplant animal cells and organs into patients—"xenotransplantation"—steadily mounts.

Blending Species

The thought of combining parts from different species is not at all new. Greek lore of more than 3,000 years ago featured centaurs—creatures that were half man, half horse—and the Chimera, a combination of lion, goat and serpent. As early as 1682 a Russian physician reportedly repaired the skull of a wounded nobleman using bone from a dog. But it was not until after the turn of the 20th century that doctors attempted with some regularity to graft tissues from animals into humans. For instance, in 1905 a French surgeon inserted slices of rabbit kidney into a child suffering from kidney failure. "The immediate results," he wrote, "were excellent." Nevertheless, the child died about two weeks later.

During the next two decades, several other doctors tried to transplant organs from pigs, goats, lambs and monkeys into various patients. These grafts all soon failed, for reasons that seemed puzzling at the time. Before the pioneering investigations of Nobel laureate Sir Peter Medawar at the University of London during the 1940s, physicians had little inkling of the immunologic basis of rejection.

So, with only failures to show, most doctors lost interest in transplantation. But some medical researchers persevered, and in 1954 Joseph E. Murray and his colleagues at Peter Bent Brigham Hospital in Boston performed the first truly successful kidney transplant. They avoided immunologic rejection by transplanting a kidney between identical twin brothers (whose organs were indistinguishable to their immune systems). Subsequently, Murray and others were able to transplant kidneys from more distantly related siblings and, finally, from unrelated donors, by administering drugs to suppress the recipient's innate immune response.

Medical practice has since grown to include transplantation of the heart, lung, liver and pancreas. But these accomplishments have brought tragedy with them: because of the shortage of donated organs, most people in need cannot be offered treatment. Of the tens of thousands of patients in the U.S. every year deemed good candidates for a transplant, less than half receive a donated organ. The shortfall will become even more dire once doctors perfect methods to treat diabetes by transplanting pancreatic islet cells, which produce insulin. Islet replacement is simpler than transplanting the whole pancreas, but it may require harvesting cells from several donors to treat each patient.

Fortunately, scientists did not entirely abandon the possibility of using animal tissues in patients after human organ transplants came into vogue. During the 1960s, medical researchers continued to investigate exactly why organs transplanted between widely different species fail so rapidly. A major cause, they learned, is that the recipient's blood harbors antibody molecules that bind to the donated tissues. (These antibodies are normally directed against infectious microbes but can also respond to components of transplanted organs.) The attachment of these antibodies then activates special "complement" proteins in the blood, which in turn trigger destruction of the graft.

Such hyperacute rejection of foreign tissue—which begins within minutes or, at most, hours after the surgery—destroys the capillaries in the transplanted organ, causing it to hemorrhage massively. Although this reaction presents an imposing barrier to xenotransplantation, recent experiments suggest that scientists may yet overcome it.

For example, in 1992 David J. G. White and his colleagues at the University of Cambridge managed to create "transgenic" pigs, bearing on the inner walls of their blood vessels proteins that can prevent human complement proteins from doing damage. They did this by introducing into pig embryos a human gene that directs the production of a human complement-inhibiting protein [see "Transgenic Livestock as Drug Factories," page 570]. White and his co-workers have not yet tested how tissues from these pigs fare in a human host, but organs from such genetically engineered pigs have functioned for as long as two months in monkeys, because the pig cells that are in direct contact with the host's immune system are able to quash the first wave of attack.

Other methods may also serve to thwart hyperacute rejection. In 1991 one of us (Cooper), along with several other investigators, identified the specific molecular fragments, or antigens, on pig tissues that human antibodies target. The cells lining pig vasculature have on their surfaces antigens made up of a particular sugar group. So it may be possible to breed (or indeed to clone) a line of genetically engineered pigs that lack this troublesome sugar group.

Yet another strategy to prevent hyperacute rejection would be to alter the recipient's immune system so that it cannot destroy the transplanted tissue. For example, using standard apparatus, doctors can remove from the patient's blood all the antibodies to pig tissue. It is also possible to deplete the complement proteins temporarily or otherwise interfere with their activation. Remarkably, animal studies suggest that if surgeons transplant a pig organ while the patient's immune system is so suppressed, the organ may—for reasons that remain largely mysterious—achieve accommodation, a state that enables it to survive even after the host's antibodies and complement return to normal levels. The transplanted organ then continues to work despite a distinct lack of tolerance from the host's immune system.

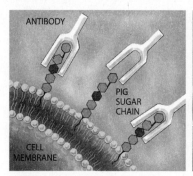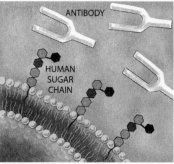

HYPERACUTE REJECTION of a pig organ transplanted into a patient would very likely occur in minutes. It ensues after antibodies bind to the linear sugar chains lining pig blood vessels (*left*). But tissues from pigs genetically engineered to carry the angular sugar groups found in people with type O blood should not elicit such reactions (*right*).

Unfortunately, researchers have not yet managed to induce accommodation reliably in animals undergoing xenotransplantation. But Guy Alexandre and his colleagues at the University of Louvain Medical School in Belgium have achieved it in certain patients who received human organs from donors with incompatible blood types—a situation that, like xenotransplantation, normally sparks hyperacute rejection.

Fostering Tolerance

Investigators studying xenotransplantation are optimistic that, with some combination of these methods, immediately harmful immune reactions can be overcome. Yet grafts of animal tissues in patients would still fall prey to more delayed forms of immune rejection, which can take days or weeks to develop. In particular, the so-called cellular immune response to grafts from animals is likely to be at least as strong as the robust attacks that white blood cells of the immune system often mount against organs transplanted from one person to another.

Avoiding such delayed reactions might require massive doses of immunosuppressive drugs, such as cyclosporine, to be given indefinitely, and the risks of toxicity, infections and other complications would be excessive.

Newly devised immunosuppressive agents should help, but it would clearly be more desirable if the human body could be induced to accept animal tissues without requiring ongoing drug therapy. That happy condition might seem impossibly difficult to arrange. But hope springs from the observation that long-term organ acceptance, or immunologic tolerance, has occurred spontaneously in a few people who have received human organs. Doctors of these patients were able to reduce, and ultimately eliminate, the normal regimen of immunosuppressive drugs.

Though still an elusive goal, the induction of immunologic tolerance is an area of vigorous research, and advances are sure to come. Curiously, it may ultimately prove easier to achieve tolerance with xenotransplantation than with traditional organ transplants. Donated human organs need to be procured urgently under emergency conditions, but animal organs would be available on demand. That flexibility might give physicians adequate time to reprogram the immune system of the recipient.

One way to create tolerance involves modifying the immune system of the patient with bone marrow cells from the donor animal. (Bone marrow is the source of all components of the blood, including the white blood cells of the immune system.) Once introduced, the donated cells spread and mature, creating a "chimeric" immune system that is part donor, part recipient. The aim is to alter the patient's immune system so that it does not recognize as foreign either the donated cells or subsequently transplanted tissues from the same animal.

Following this strategy, David H. Sachs and his colleagues at Massachusetts General Hospital injected bone marrow cells from donor pigs (along with substances to stimulate proliferation of the cells) into baboons. These animals had undergone a course of radiation to deplete their immune systems

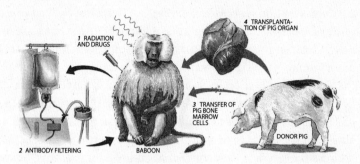

1 RADIATION AND DRUGS

4 TRANSPLANTA-TION OF PIG ORGAN

3 TRANSFER OF PIG BONE MARROW CELLS

DONOR PIG

2 ANTIBODY FILTERING

BABOON

PREVENTING DELAYED REJECTION of cross-species organ transplants might be possible by altering the recipient's immune system so that it includes components from the donor. To test this strategy in animals, a baboon slated to receive bone marrow cells (the source of all immune cells) from a pig was first given radiation and drugs (*1*) to prevent immune rejection of the transplanted cells. The baboon's blood was also filtered (*2*) to remove antibody molecules that would react with pig cells. Finally, the baboon received the bone marrow cells from the donor pig (*3*). Afterward, killer *T* cells isolated from the baboon's immune system did not attack cells from the donor pig. If other components of the recipient's immune system could be equally tamed, organs transplanted from a donor pig (*4*) should be able to survive indefinitely in the new host.

temporarily and prevent rejection of the pig bone marrow cells. The researchers also filtered from the blood of the baboons those antibodies directed against pig tissues and administered a brief course of immunosuppressive drugs. Although the baboons' immune systems eventually killed most of the transplanted cells, some pig DNA survived in one of the baboons for almost a year. What is more, an important component of this chimeric baboon's immune system—the aggressive killer T cells—no longer reacted to the pig cells as foreign.

Such research may yield ways to prevent immune rejection of organs transplanted from animals, but truly effective measures are probably still some years away. Another scheme for evading rejection is, however, already undergoing clinical trials: immunoisolation. Following this approach, physicians physically sequester transplanted tissue within a membrane that allows small molecules (such as nutrients, oxygen and certain therapeutic agents) to cross it while blocking large molecules (such as antibodies) and white blood cells from reaching the graft. This tactic is feasible only for protecting isolated cells or small packages of tissues, not for whole organs. So it does not address the needs of someone who requires, for example, a new heart or kidney. It should nonetheless be valuable for treating many disorders. And it offers some practical advantages: physicians can manipulate cells or small masses of tissue comparatively easily and can maintain them outside the body for longer periods than are possible when working with intact organs.

Recent attempts at using encapsulated cells from animals to treat liver failure, chronic pain and amyotrophic lateral sclerosis (Lou Gehrig's disease) have all shown promise in clinical trials. Medical researchers may soon try to implant immunoisolated cells from animals to provide the blood-clotting factors hemophiliacs need or to produce nerve growth factors that might help reverse certain neurodegenerative disorders.

Some investigators are especially eager to treat diabetes with isolated pig islet cells. Although one of us (Chick) pioneered the use of "perfused" devices (large sheathed implants connected to a supply of blood) for this purpose, it is easy to see some disadvantages to that particular technique. Most important, the patient requires major surgery, and the device is apt to become clotted. Engineering hollow plastic fibers or chambers unconnected to the bloodstream to isolate cells from the recipient's immune system also has drawbacks: although the surgery needed would be less traumatic than for a perfused device, it is unclear how well a patient could tolerate the plastic materials or having the implant replaced many times—a likely requirement of long-term therapy.

In an effort to overcome these difficulties, two of us (Lanza and Chick), along with colleagues at BioHybrid Technologies, have developed ways to encase cells in small, biodegradable capsules that can be injected under the skin or placed in the abdominal cavity with a syringe. Less than a gram of encapsulated islets from pigs should supply a diabetic patient with normal amounts of insulin. Although a vast number of cells are involved, the total volume required for these implants would be only a few dozen cubic centimeters.

It may take many years before physicians can routinely outwit evolution—as some have labeled the goal of xenotransplantation—and replace any failing organ with an animal substitute. But the transplantation of isolated cells and tissues appears poised on the threshold of modern medical practice. And we are optimistic that soon there will be some true successes to report.

Animals, plants, and specific cells have all been successfully cloned. The next step is obvious: Clone a human being.

Right now, the technology is in place that could make such a science fiction scenario a reality. With such a powerful tool in our hands myriad ethical questions come into play: Should people clone themselves? Should genetic diseases be screened out from the human family? What would the impact be on cloned children?

The answer to questions like these and many others could set the course for the future of this young science.

had cloned an adult sheep
ediately captivated the pub
some cautious researchers
ther the success might not
the pan. After all, the Sco
ers had to try 277 times b
ceeded in producing the clo
ly. Unless the efficiency o
ning process could be great

I, Clone

Ronald M. Green

Within the first five years of the next century, a team of scientists somewhere in the world will probably announce the birth of the first cloned human baby. Like Louise Brown, the first child born as the result of in vitro fertilization 21 years ago, the cloned infant will be showered with media attention. But within a few years it will be just one of hundreds or thousands of such children around the world.

It has been possible to envision such a scenario realistically only since Ian Wilmut and his colleagues at the Roslin Institute near Edinburgh, Scotland, announced in February 1997 that they had cloned a sheep named Dolly from the udder cells of a ewe. The technique used by Wilmut and his co-workers—a technology called somatic-cell nuclear transfer—will probably be the way in which the first human clone will be created [see "A Clone in Sheep's Clothing," page 37].

Research on the basic processes of cell differentiation holds out the promise of dramatic new medical interventions and cures. Burn victims or those with spinal cord injuries might be provided with replacement skin or nerve tissue grown from their own body cells. The damage done by degenerative disorders such as diabetes, Parkinson's disease or Alzheimer's disease might be reversed. In the more distant future, scientists might be able to grow whole replacement organs that our bodies will not reject [see "Mother Nature's Menders," page 69].

In view of the still unknown physical risks that cloning might impose on the unborn child, caution is appropriate. Of the 29 early embryos created by somatic-cell nuclear transfer and implanted into various ewes by Roslin researchers, only one, Dolly, survived, suggesting that the technique currently has a high rate of embryonic and fetal loss. Dolly herself appears to be a normal three-year-old sheep—she recently gave birth to triplets following her second pregnancy. But a recent report that her telomeres—the tips of chromosomes, which tend to shrink as cells grow older—are shorter than normal for her age suggests that her life span might be reduced. This and other matters must be sorted out and substantial further animal research will need to be completed before cloning can be applied safely to humans.

Eventually animal research may indicate that human cloning can be done at no greater physical risk to the child than intro-vertal fertilization posed when it was first introduced. One would hope that such research will be done openly in the U.S., Canada, Europe or Japan, where established government agencies exist to provide careful oversight of the implications of the studies for human subjects. Less desirably, but more probably, it might happen in clandestine fashion in some offshore laboratory where a couple desperate for a child has put their hopes in the hands of a researcher seeking instant renown.

Given the pace of events, it is possible that this researcher is already at work. For now, the technical limiting factor is the availability of a sufficient number of ripe human eggs. If Dolly is an indication, hundreds might be needed to produce only a few viable cloned embryos. Current assisted-reproduction regimens that use hormone injections to induce egg maturation produce at best only a few eggs during each female menstrual cycle. But scientists might soon resolve this problem by improving ways to store frozen eggs and by developing methods for inducing the maturation of eggs in egg follicles maintained in laboratory culture dishes.

Who First?

Once human cloning is possible, why would anyone want to have a child that way? As we consider this question, we should put aside the nightmare scenarios much talked about in the press. These include dictators using cloning to amass an army of "perfect soldiers" or wealthy egotists seeking to produce hundreds or thousands of copies of themselves. Popular films such as *Multiplicity* feed these nightmares by obscuring the fact that cloning cannot instantaneously yield a copy of an existing adult human being. What somatic-cell nuclear transfer technology produces are cloned human embryos. These require the labor- and time-intensive processes of gestation and child rearing to reach adulthood. Saddam Hussein would have to wait 20 years to realize his dream of a perfect army. And the Donald Trumps of the world would also have to enlist thousands of women to be the mothers of their clones.

For all their efforts, those seeking to mass-produce children in this way, as well as others who seek an exact copy of someone else, would almost certainly be disappointed in the end. Although genes contribute to the array of abilities and limits each of us possesses, from conception forward their expression is constantly shaped by environmental factors, by the unique experiences of each individual and by purely chance factors in biological and social development. Even identical twins (natural human clones) show different physical and mental characteristics to some degree. How much more will this be true of cloned children raised at different times and in different environments from their nucleus-donor "parent"? As one wit has observed, someone trying to clone a future Adolf Hitler might instead produce a modestly talented painter.

So who is most likely to want or use human cloning? First are those individuals or couples who lack the gametes (eggs or sperm) needed for sexual reproduction. Since the birth of Louise Brown, assisted-reproduction technologies have made remarkable progress in helping infertile women and men become parents. Women with blocked or missing fallopian tubes, which carry the eggs from the ovaries to the womb, can now use in vitro fertilization to overcome the problem, and those without a functional uterus can seek the aid of a surrogate mother. A male who produces too few viable sperm cells can become a father using the new technique of intracytoplasmic sperm injection, which involves inserting a single sperm or the progenitor of a sperm cell into a recipient egg.

Despite this progress, however, women who lack ovaries altogether and men whose testicles have failed to develop or have been removed must still use donor gametes if they wish to have a child, which means that the child will not carry any of their genes. Some of these individuals might prefer to use cloning technology to have a genetically related child. If a male totally lacks sperm or the testicular cells that make it, a nucleus from one of his body cells could be inserted into an egg from his mate that had had its nucleus removed. The child she would bear would be an identical twin of its father. For the couple's second child, the mother's nucleus could be used in the same procedure.

One very large category of such users of cloning might be lesbian couples. Currently if two lesbians wish to have a child, they must use donor sperm. In an era of changing laws about the rights of gamete donors, this opens their relationship to possible intervention by the sperm donor if he decides he wants to play a role in raising the child. Cloning technology avoids this problem by permitting each member of the pair to bear a child whose genes are provided by her partner. Because the egg-donor mother also supplies to each embryo a small number of mitochondria—tiny energy factories within cells that have some of their own genetic material—this approach even affords lesbian couples an approximation of sexual reproduction. (Cloning might not be used as widely by gay males, because they would need to find an egg donor and a surrogate mother.)

A second broad class of possible users of cloning technologies includes individuals or couples whose genes carry mutations that might cause serious genetic disease in their offspring. At present, if such people want a child with some genetic relationship to themselves, they can substitute donated sperm or eggs for one parent's or have each embryo analyzed genetically using preimplantation genetic diagnosis so that only those embryos shown to be free of the disease-causing gene are transferred to the mother's womb. The large number of genetic mutations contributing to some disorders and the uncertainty about which gene mutations cause some conditions limit this approach, however.

Some couples with genetic disease in their families will choose cloning as a way of avoiding what they regard as "reproductive roulette." Although the cloned child will carry the same problem genes as the parent who donates the nucleus, he or she will in all likelihood enjoy the parent's state of health and will be free of the additional risks caused by mixing both parents' genes during sexual reproduction. It is true, of course, that sex is nature's way of developing new combinations of genes that are able to resist unknown health threats in the future. Therefore, cloning should never be allowed to become so common that it reduces the overall diversity in the human gene pool. Only a relatively few couples are likely to use cloning in this way, however, and these couples will reasonably forgo the general advantages conveyed by sexual reproduction to reduce the immediate risks of passing on a genetic disease to their child.

Cloning also brings hope to families with inherited genetic diseases by opening the way to gene therapy. Such therapy—the actual correction or replacement of defective gene sequences in the embryo or the adult—is the holy grail of genetic medicine. To date, however, this research has been slowed by the inefficiency of the viruses that are now used as vectors to carry new genes into cells. By whatever means they are infused into the body, such vectors seem to reach and alter the DNA in only a frustratingly small number of cells.

Cloning promises an end run around this problem. With a large population of cells from one parent or from an embryo created from both parents' gametes, vectors could be created to convey the desired gene sequence. Scientists could determine which cells have taken up the correct sequence using fluorescent tags that cause those cells to glow. The nucleus of one of these cells could then be inserted into an egg whose own nucleus has been removed, and the "cloned" embryo could be transferred to the mother's womb. The resulting child and its descendants would thereafter carry the corrected gene in every cell of their bodies. In this way, age-old genetic maladies such as Tay-Sachs disease, cystic fibrosis, muscular dystrophy or Huntington's disease could be eliminated completely from family trees.

Cloning and Identity

Merely mentioning these beneficial uses of cloning raises difficult ethical questions. The bright hope of gene therapy is dimmed somewhat by the reawakening of eugenic fears. If we can manipulate embryos to prevent disease, why not go further and seek "enhancements" of human abilities? Greater disease resistance, strength and intelligence all beckon alluringly, but questions abound. Will we be tampering with the diversity that has been the mainstay of human survival in the past? Who will choose the alleged enhancements, and what will prevent a repetition of the terrible racist and coercive eugenic programs of the past?

Even if it proves physically safe for the resulting children, human cloning raises its own share of ethics dilemmas. Many wonder, for example, about the psychological well-being of a cloned child. What does it mean in terms of intrafamily relations for someone to be born the identical twin of his or her parent? What pressures will a cloned child experience if, from his or her birth onward, he or she is constantly being compared to an esteemed or beloved person who has already lived? The problem may be more acute if parents seek to replace a deceased child with a cloned replica. Is there, as some ethicists have argued, a "right to one's unique genotype," or genetic code—a right that cloning violates? Will cloning lead to even more serious violations of human dignity? Some fear that people may use cloning to produce a subordinate class of humans created as tissue or organ donors.

Some of these fears are less substantial than others. Existing laws and institutions should protect people produced by cloning from exploitation. Cloned humans could no more be "harvested" for their organs than people can be today. The more subtle psychological and familial harms are a worry, but they are not unique to cloning. Parents have always imposed unrealistic expectations on their children, and in the wake of widespread divorce and remarriage we have grown familiar with unusual family structures and relationships. Clearly, the initial efforts at human cloning will require good counseling for the parents and careful follow-up of the children. What is needed is caution, not necessarily prohibition.

As we think about these concerns, it is useful to keep a few things in mind. First, cloning will probably not be a widely employed reproductive technology. For many reasons, the vast majority of heterosexuals will still prefer the "old-fashioned," sexual way of producing children. No other method better expresses the loving union of a man and a woman seeking to make a baby.

Second, as we think about those who would use cloning, we would do well to remember that the single most important factor affecting the quality of a child's life is the love and devotion he or she receives from parents, not the methods or circumstances of the person's birth. Because children produced by cloning will probably be extremely wanted children, there is no reason to think that with good counseling support for their parents they will not experience the love and care they deserve.

What will life be like for the first generation of cloned children? Being at the center of scientific and popular attention will not be easy for them. They and their parents will also have to negotiate the worrisome problems created by genetic identity and unavoidable expectations. But with all these difficulties, there may also be some novel satisfactions. As cross-generational twins, a cloned child and his or her parent may experience some of the unique intimacy now shared by sibling twins. Indeed, it would not be surprising if, in the more distant future, some cloned individuals chose to perpetuate a family "tradition" by having a cloned child themselves when they decide to reproduce.

Outside of the macroworld of human beings and animals, cloning also applies to the microworld of synthetic molecules. A series of ongoing investigations have established that many man-made molecules can be designed to clone themselves.

The success of these efforts, cataloged in this article, could bring a new understanding about the mechanics of how life began on Earth.

Synthetic Self-Replicating Molecules

Julius Rebek, Jr.

Imagine a molecule that likes its own shape: finding a copy of itself, it will join neatly with its twin, forming for a while a complete entity. If the original molecule is presented with the component parts of itself, it will assemble these into additional replicas. The process will continue as long as the supply of components lasts.

My colleagues and I at the Massachusetts Institute of Technology have designed such self-assembling molecules and crafted them in the laboratory. Our efforts are intended to illuminate the ways in which life might have arisen. Probably it began when molecules came into existence that were capable of reproducing themselves. Our organic molecules, although they operate outside of living systems, help to elucidate some of the essential principles of self-replication.

Attempts to imitate life are still very young compared to the beginnings of life on the earth perhaps three and a half billion years ago. No one can say for sure what the atmospheric or terrestrial conditions were at the time or which molecule crossed the critical frontier between organic chemistry and biology.

In 1953 Stanley L. Miller, then at the University of Chicago, made one of the first—if not the first—attempts to recreate this transition. He formed amino acids in a mixture of water, methane, ammonia and hydrogen—substances thought to have been present on the primitive earth—by subjecting the chemicals to an electric discharge. James D. Watson and Francis Crick's unraveling of the structure of DNA—also in 1953—further fueled this quest.

For some decades now, the most widely accepted recipe for the origin of life has specified DNA or RNA in lukewarm water, neither acidic nor alkaline, plus only those reagents that were presumably part of the primitive earth's atmosphere. Practitioners of prebiotic chemistry such as Miller and Leslie E. Orgel of the Salk Institute for Biological Studies in San Diego have provided deep insights into molecular replication under these constraints. Indeed, countless metric tons of DNA replicate every day in living creatures in precisely this manner, under the surveillance of a host of enzymes.

But recent findings indicate that the primitive earth was perhaps not quite as hospitable as the "warm pond" paradigm would have. Such suspicions, coupled with discoveries of organisms living at (literally) blood-curdling temperatures or near sulfurous volcanic vents at the bottom of the ocean, have led scientists to suggest that life possibly arose by some very different route. Perhaps the only qualities essential to the molecules that become life are detailed in two remarkable books by Richard Dawkins, *The Blind Watchmaker* and *The Selfish Gene*. Although written 15 years ago, Dawkins's sketches astonishingly prefigure the results of the past four years of my work on self-replication.

Molecules, natural or synthetic, are able to replicate when their shapes and chemistry have a feature called complementarity. By virtue of the way a molecule occupies space and the way its attracting atoms or groups of atoms are distributed along its arms, one molecule may fit snugly into the chemical nooks and crannies of another. The "goodness of fit" between two such complementary molecules thus depends not only on their spatial structure but also on the different kinds of chemical bonds that hold them together in groups. Such groups, or "complexes," form and dissipate rapidly in microseconds or nanoseconds—times that are very short, yet long enough for chemical reactions to take place.

The forces holding complexes together are many times weaker than the covalent bonds binding atoms into molecules. One kind of force—important in complexes—is called a hydrogen bond. This bond comes about when a hydrogen atom possessing a partial positive charge is attracted to, for instance, an oxygen atom that has a partial negative charge. More general attractions of this class go by the name of polar interactions.

Another kind of force, the van der Waals force, is more subtle: if correctly positioned, electrons of one molecule can jostle away those of another, creating a charge imbalance that results in attraction. Yet a third kind of attraction is "aromatic stacking"—an arrangement that flat organic molecules (often having a pleasant odor, hence the name) sometimes assume when they do not like the solvent they find themselves in. By sidling up to one another, flat surface to flat surface, they can squeeze out all the solvent molecules between them and achieve a more stable, stacked configuration.

Once a complex forms, the molecular surfaces that match up with one another are relatively protected. Destructive solvents, dissolved acids, bases or oxidants cannot get to them. Strong covalent bonds then have time to join the complementary parts. Sometimes two of three molecules in a complex link together; the third merely serves to ease the process. Such a coupling gives rise to a rather popular scheme for replication—the one preferred by DNA. A simple depiction of this scheme uses concave and convex shapes. A concave molecular surface—lined with appropriately enticing atoms—can recognize and surround its convex complement. Further, it can act as a

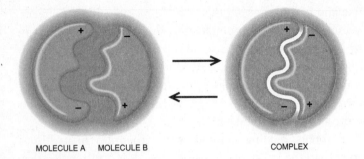

MOLECULE A   MOLECULE B   COMPLEX

MOLECULAR RECOGNITION occurs when two fragments whose geometric and chemical properties complement one another form a complex. The + and − signs indicate electrostatic attractions. Moreover, the solvent is squeezed out between the molecules, helping to stabilize the short-lived complex.

mold for assembling the convex molecule from its component parts. In turn, the convex molecule serves as a template for gathering and fusing the component parts of the concave one. These two replication events—each molecule forming the other—establish what is called a bi-cycle. Our recent experiments indicate that a bi-cycle can be extremely efficient.

There is an alternative paradigm of replication: two complementary molecules in a complex can join at some site that is not on the recognition surface. They form a single molecule, one end of which is complementary to the other—and the whole is complementary to itself. The recognition surfaces at the ends of this new, self-complementary molecule are still accessible to other molecules. The ends can each gather a fragment identical to that at the other end.

Once gathered, the two new components cannot move freely and travel through space in tandem; the chances of their becoming linked to each other are greatly enhanced. Thus, the self-complementary entity makes a copy—and in similar manner, many copies—of itself. No enzymes are needed: the molecule catalyzes its own formation.

This is the method we have used in the laboratory to make

molecules capable of reacting with one another in ways reminiscent of life. Among them are molecules that bear a passing resemblance to genetic materials—specifically, to nucleic acid components known as adenines. Adenines are flat; besides, they have hydrogen and nitrogen atoms that can form hydrogen bonds with the oxygen and hydrogen atoms of their complementary molecules, called imides. Our imides are constructed from a humpbacked molecule, Kemp's triacid, the skeleton of which folds over in such a way that large, concave structures can easily be fashioned from it. So the imide features a hydrogen bonding site crookedly attached to an aromatic stacking surface; these fit perfectly with the hydrogen bonding site and the flat stacking surface of adenine.

When associated together in a complex, the adenine and the imide become covalently attached, forming a self-complementary molecule. Our early attempts to get this molecule to self-replicate were thwarted by its unforeseen floppiness. Although some flexibility is helpful for molecular recognition—a leather boot is easier to slip on than a wooden one—a lot of flexibility can make fitting very difficult—try slipping on a sock without using your hands.

Happily, this situation was curable. The remedy called for inserting a larger and more rigid molecule in place of the single chain to prevent folding. Our choice was a larger stacking surface, a naphthalene, bolstered by a less flexible link between the two components, a cyclic ribose group.

This new J-shaped molecule, adenine ribose naphthalene imide (ARNI for short), provided us with our first instance of replication. Using high-performance liquid chromatography to detect minute changes in chemical concentrations, Tjivikua Tjama, my graduate student from Namibia, and Pablo Ballester, a postdoctoral visitor from Majorca, achieved the result. They compared the rate of formation of ARNI in a solution that contained only its components with the rate of formation when some ARNI was added. The presence of ARNI

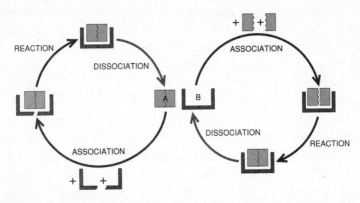

REPLICATION BI-CYCLE involves two molecules of complementary shapes, represented by block A and sleeve B, into which it fits. In the left cycle, the block (*middle*) collects the two parts of the sleeve (*bottom*) around it to form a complex (*left*); the parts then react to form a whole sleeve (*top*). The block and sleeve quickly dissociate. In the right cycle, it is the sleeve that assembles the fragments of the block. Thus, the two complementary molecules catalyze each other's formation.

increased the rate of formation, clear evidence of the presence of a self-replicating system.

If one plots the progress of a reaction through time, one generally derives a curve that assumes the shape of a reclining parabola. The product forms fastest at the beginning, when the reactants are at their highest concentrations; the rate of formation slows down as the reactants are consumed. For an autocatalytic reaction—one in which the product, like our ARNI, catalyzes its own formation—the growth curve should be S-shaped, or "sigmoidal." The reaction begins slowly. As the product appears and begins to act as a catalyst, the reaction accelerates. An upward curve results. Finally, as the materials are consumed, the reaction grinds to a halt.

The degree of sigmoidal curvature depends on several factors, the most important of which is the efficiency of the autocatalytic step. If the background reaction—in which the components combine by themselves, without getting help from the self-replicating molecule—is too strong, it can swamp the signal from the self-replicating process. In 1990 Günther von Kiedrowski and his co-workers in Göttingen showed that a self-replicating nucleic acid could exhibit such sigmoidal growth—proving that the autocatalyzed synthesis is in this case more efficient than the random one.

Although ARNI did not show sigmoidal growth, our next attempt, ARBI, did. We slowed the background reaction rates by giving ARBI a slightly longer stacking element, a biphenyl instead of naphthalene. We now had proof of a bona fide synthetic self-replicating molecule.

Is it alive? Not by most current definitions. Our (or its) critics were quick to point out that as a life-form, ARBI had severe limitations: the molecule would make copies only of itself. To allow evolution, a self-replicating molecule has to be capable of "making mistakes": occasionally synthesizing other molecules that can perhaps be better replicators. Unlike art and music critics, those in science at least indicate in which direction improvements may lie. We responded by devising molecules that were capable of making—indeed, that were incapable of not making—mistakes.

In organic chemistry, a "mistake" is made when there is a lack of selectivity between reaction partners. We needed a molecule that would catalyze not only its own formation but also that of a molecule of similar shape. Besides, at least one of these two molecules had to be able to change into a more efficient replicator.

Although mutation is considered to be the driving force for most evolutionary alterations, another significant paradigm for change is recombination. Two chromosomes can split, exchange strings of DNA and rejoin, thus combining their characteristics. Also, certain computer programs attempt to "teach" strings of information to solve a problem. If the strings are allowed to split and recombine at random, they soon give rise to much better problem solvers. Mutation allows for single, small changes;

recombination, on the other hand, allows the creation of hybrids that are very different from the progenitors.

Our interest in demonstrating recombination at the molecular level led us to develop an entirely new set of self-replicating molecules. The principle was the same: two complementary molecules were joined by a covalent bond to give a single, self-complementary whole that could aid its own synthesis. Qing Feng and another student, Tae Kwo Park, devised a replicating system based on a different component of nucleic acids, thymine. Some time earlier Park had developed a synthetic receptor that would recognize thymine's imide nucleus and also lie on thymine's flat aromatic surface. This receptor featured a U-shaped molecular skeleton. The bottom of the U was a large, rigid aromatic spacer known as xanthene; one arm of the U featured an amine and the other arm a diaminotriazine, the receptor for thymine. When the latter two became joined by a covalent bond, a self-complementary unit was generated, called diaminotriazine xanthene thymine, or DIXT. We were able to show that DIXT was also self-replicating.

The stage was now set for a recombination experiment. Could the adenine-based replicators and the thymine-based replicators, when placed in the same vessel, shuffle their components into new combinations? They did indeed. Even so, we were surprised by the results. One of the new recombinants, ART (adenine ribose thymine), was the most prolific replicator we had yet encountered, whereas the other one, DIXBI (diaminotriazine xanthene biphenyl imide), was unable to replicate at all—it was "sterile."

How did this difference in the ability to replicate come about? The effciency of the ART replicator is easily rationalized. ART looks a good deal like a piece of DNA, possibly the best replicator in existence. Furthermore, its ribose piece makes the recognition surfaces parallel to one another, a very helpful configuration. This and the high affinity of adenine for its complement thymine make for an easily assembled complex—the intermediate stage in replication.

With this experiment we were able to show that a relatively small pool of components can give rise to a "family tree" of replicators. Three of these are effective at self-replication, but one branch of the tree dies out. To push this analogy further, it would be appropriate for the sterile molecule to be chopped up and converted into pieces that the effective replicators could use for themselves. We have made some progress in this direction. It requires equipping our molecules with acids and bases that can manipulate other molecules more actively than simple recognition will allow.

Although it has been enjoyable to pursue replication and even evolution with synthetic molecules, we have been looking to the next step in expressing life as a series of molecular reactions. We feel, as do other workers, that a key attribute of life is a boundary: a container or a cell wall that separates inside from outside and prevents desirable molecules from floating away—while keeping undesirable ones at bay.

Inspired by a naturally occurring membrane, we have made some small, initial steps toward this goal. Viruses use a protein shell as a container; the shell is made up of many identical copies of a single protein unit. The units are also self-complementary—but the recognition surfaces are oriented so that they automatically assemble into a closed shell. Indeed, Crick had predicted that many identical copies of proteins would compose the viral coat, since there is not enough information in the viral genome for many different molecules to be involved.

When we used self-complementarity as our guide, a minimalist design struck us, based on the structure of a tennis ball. Cut along its seam, a tennis ball gives two identical pieces, the convex ends of which are complementary in shape to the concave middles. René Wyler, a Swiss postdoctoral fellow, has now synthesized a structure that mimics the shape of the tennis-ball pieces, while adding chemical complementarity. The units fit together with hydrogen bonds along the seam.

There are good indications that a smaller molecule, such as a solvent chloroform molecule, can fit within our molecular tennis ball. But it is too small to accommodate even our most minimal replicators. We are now working with Javier de Mendoza of the Free University of Madrid to develop a larger molecule—a softball—that may have an interior roomy enough to hold some of our replicators.

Once we have made it past the problem of containment, the biggest obstacle to the molecular life agenda will be these questions: How can our fledgling organism harness energy? From sunlight or from other molecules? How can the component pieces of the replicators and their containers be replenished? These are the challenges of the next decade. Whether they are met or not, the efforts of chemists to answer them will surely provide insight into the organic chemistry of life—how it came about and how it continues to flourish.

Can a machine clone itself? Although the assumption that they can has been the core of may science fiction stories, the quest to build self-reproducing machines still falls short of human imagination. However, significant inroads both in theory and in practice have recently come to light.

Thus far, machines that can clone themselves exist only within the confines of a computer environment. How long it will be before they are able to step out of this controlled environment and into the real world? Perhaps the lag time will be far shorter than anyone anticipates.

burgh reported in February
had cloned an adult sheep.
diately captivated the publ
some cautious researchers
ther the success might not b
the pan. After all, the Scot
ers had to try 277 times be
eeded in producing the clor
y. Unless the efficiency of
ning process could be greatl

Go Forth and Replicate

Moshe Sipper and
James A. Reggia

Apples beget apples, but can machines beget machines? Today it takes an elaborate manufacturing apparatus to build even a simple machine. But could we endow an artificial device with the ability to multiply on its own? Self-replication has long been considered one of the fundamental properties separating the living from the nonliving. Historically our limited understanding of how biological reproduction works has given it an aura of mystery and made it seem unlikely that it would ever be done by a manmade object. It is reported that when René Descartes averred to Queen Christina of Sweden that animals were just another form of mechanical automata, Her Majesty pointed to a clock and said, "See to it that it produces offspring."

The problem of machine self-replication moved from philosophy into the realm of science and engineering in the late 1940s with the work of eminent mathematician and physicist John von Neumann. Some researchers have actually constructed physical replicators. Forty years ago, for example, geneticist Lionel Penrose and his son, Roger (the famous physicist), built small assemblies of plywood that exhibited a simple form of self-replication. But self-replication has proved to be so difficult that most researchers study it with the conceptual tool that von Neumann developed: two-dimensional cellular automata.

Implemented on a computer, cellular automata can simulate a huge variety of self-replicators in what amount to austere universes with different laws of physics from our own. Such models free researchers from having to worry about logistical issues such as energy and physical construction so that they can focus on the fundamental questions of information flow. How is a living being able to replicate unaided, whereas mechanical objects must be constructed by humans? How does replication at the level of an organism emerge from the numerous interactions in tissues, cells and molecules? How did Darwinian evolution give rise to self-replicating organisms?

The emerging answers have inspired the development of self-repairing silicon chips and autocatalyzing molecules [see "Synthetic Self-Replicating Molecules," page 99]. And this may be just the beginning. Researchers in the field of nanotechnology have long proposed that self-replication will be crucial to manufacturing molecular-scale machines, and proponents of space exploration see a macroscopic version of the process as a way to colonize planets using in situ materials. Recent advances have given credence to these futuristic-sounding ideas. As with other scientific disciplines, including genetics, nuclear energy and chemistry, those of us who study self-replication face the twofold challenge of creating replicating machines and avoiding dystopian predictions of devices running amok. The knowledge we gain will help us separate the wheat of good technologies from the chaff of destructive ones.

Playing Life

Science-fiction stories often depict cybernetic self-replication as a natural development of current technology, but they gloss over the profound problem it poses: how to avoid an infinite regress. A system might try to build a clone using a blueprint, that is, a self-description. Yet the self-description is part of the machine, is it not? If so, what describes the description? And what describes the description of the description? Self-replication in this case would be like asking an architect to make a perfect blueprint of his or her own studio. The blueprint would have to contain a miniature version of the blueprint, which would contain a miniature version of the blueprint and so on. Without this information, a construction crew would be unable to re-create the studio fully; there would be a blank space where the blueprint had been.

Von Neumann's great insight was an explanation of how to break out of the infinite regress. He realized that the self-description could be used in two distinct ways: first, as the instructions whose interpretation leads to the construction of an identical copy of the device; next, as data to be copied, uninterpreted, and attached to the newly created child so that it, too, possesses the ability to self-replicate. With this two-step process, the self-description need not contain a description of itself. In the architectural analogy, the blueprint would include a plan for building a photocopy machine. Once the new studio and the photocopier were built, the construction crew would simply run off a copy of the blueprint and put it into the new studio.

Living cells use their self-description, which biologists call the genotype, in exactly these two ways: transcription and translation. Von Neumann made this transcription-translation distinction several years before molecular biologists did, and his work has been crucial in understanding self-replication in nature.

To prove these ideas, von Neumann and mathematician colleague Stanislaw M. Ulam came up with the idea of cellular automata. A cellular-automata simulation involves a chessboardlike grid of squares, or cells, each of which is either empty or occupied by one of several possible components. At discrete intervals of time, each cell looks at itself and its neighbors and decides whether to metamorphose into a different component. In making this decision, the cell follows relatively simple rules, which are the same for all cells. These rules constitute the fundamental physics of the cellular-automata world. All decisions and actions take place locally; cells do not know directly what is happening outside their immediate neighborhood.

The apparent simplicity of cellular automata is deceptive; it does not imply ease of design or poverty of behavior. The most famous cellular automata, John Horton Conway's Game of Life, produces amazingly intricate patterns. Many questions about the dynamic behavior of cellular automata are formally unsolvable. The outcomes cannot be deduced from the local rules; to see what pattern will actually be produced, you need to simulate it fully. In its own way, a cellular-automata model can be just as complex as the real world.

Copy Machines

Within cellular automata, self-replication occurs when a group of components—a "machine"—goes through a sequence of steps to construct a nearby duplicate of itself. Von Neumann's machine was based on a universal constructor, a machine that, given the appropriate instructions, could create any pattern. The constructor consisted of numerous types of components spread over tens of thousands of cells and required a book-length manuscript to be specified. It has still not been simulated in its entirety, let alone actually built, on account of its complexity. A constructor would be much more complicated in the Game of Life because the functions performed by single cells in von Neumann's model—such as transmission of signals and generation of new components—would have to be performed by composite structures in Life.

Going to the other extreme, it is easy to find trivial examples of self-replication. For example, suppose a cellular automata has only one type of component, labeled +, and that each cell follows only a single rule based on the contents of cells to the left, to the right, above and below: if exactly one of the four neighboring cells contains a +, then the cell becomes a +; otherwise it becomes vacant. With this rule, a single + grows into four more +'s, each of which grows likewise, and so forth.

Such weedlike proliferation does not shed much light on the principles of replication, because there is no significant machine. Of course, that invites the question of how you would tell a "significant" machine from a trivially prolific automata. No one has yet devised a satisfactory answer. What is clear, however, is that the replicating structure must in some sense be complex. For example, it must consist of multiple, differing components whose interactions collectively bring about replication—the proverbial "whole must be greater than the sum of the parts." The existence of multiple distinct components permits a self-description to be stored within the replicating structure.

In the years since von Neumann's seminal work, many researchers have probed the domain between the complex and the trivial, developing replicators that require fewer components, less space or simpler rules. A major step forward was taken in 1984 when Christopher G. Langton, then at the University of Michigan, observed that looplike storage devices—which had formed modules of earlier self-replicating machines—could be programmed to replicate on their own. These devices typically consist of two pieces: the loop itself, which is a string of components that circulate around a rectangle, and a construction arm, which protrudes from a corner of the rectangle into the surrounding space. The circulating components constitute a recipe for the loop—for example, "go three squares ahead, then turn left." When this recipe reaches the construction arm, the automata rules make a copy of it. One copy continues around the loop, the other goes down the arm, where it is interpreted as an instruction.

By giving up the requirement of universal construction, which was central to von Neumann's approach, Langton showed that a replicator could be constructed from just seven unique components occupying only 86 cells. Even smaller and simpler self-replicating loops have been devised by one of us (Reggia) and our colleagues. Because they have multiple interacting components and include a self-description, they are not trivial. The smallest of these structures spans only six cells, demonstrating that nontrivial self-replication can be unexpectedly simple. Intriguingly, asymmetry plays an unexpected role: the rules governing replication are often simpler when the components are not rotationally symmetric.

Emergent Replication

All these self-replicating structures have been designed through ingenuity and much trial and error. This process is arduous and often frustrating; a small change to one of the rules results in an entirely different global behavior, most likely the disintegration of the structure in question. But recent work has gone beyond the direct-design approach. Instead of tailoring the rules to suit a particular type of structure, researchers have set up arbitrary sets of rules, filled the cellular-automata grids with a "primordial soup" of randomly selected components, and checked whether self-replicators emerged spontaneously.

In 1997 Hui-Hsien Chou and Reggia noticed that as long as the initial density of the free-floating components was above a certain threshold, small self-replicating loops reliably appeared. Loops that collided underwent annihilation, so there was an ongoing process of death as well as birth. Over time, loops proliferated, grew in size and evolved through mutations triggered by debris from past collisions. Although the automata rules were deterministic, these mutations were effectively random, because the system was complex and the components started off in random locations.

Such loops are intended as abstract machines and not as simulacra of anything biological, but it is interesting to compare them with biomolecular structures. A loop loosely resembles circular DNA in bacteria, the programmed instructions look like genes, and the construction arm acts as the enzyme that catalyzes DNA replication. More important, replicating loops illustrate how complex global behaviors can arise from simple local interactions. For example, components move around a loop even though the rules say nothing about movement; what is actually happening is that individual cells are coming alive, dying or metamorphosing in such a way that a pattern is eliminated from one position and reconstructed elsewhere—a process that we perceive as motion. In short, cellular automata act locally but appear to think globally. Much the same is true of molecular biology.

Are loops the only simple yet nontrivial self-replicators? In a recent computational experiment, Jason Lohn and Reggia experimented not with different structures but with different sets of rules. Starting with an arbitrary block of four components, they found they could determine a set of rules that made the block self-replicate. They discovered these rules via a genetic algorithm, an automated process that simulates Darwinian evolution.

The most challenging aspect of this work was the definition of the so-called fitness function—namely, the criteria by which sets of rules were judged, thus separating good solutions from bad ones and driving the evolutionary process toward rule sets that facilitated replication. The difficulty is that you cannot simply assign high fitness to those sets of rules that cause a structure to replicate, because none of the random rule sets you start off with is likely to lead to replication. The solution was to devise a fitness function composed of a weighted sum of three measures: a growth measure (the extent to which each component type generates an increasing supply of that component), a relative position measure (the extent to which neighboring components stay together) and a replicant measure (a function of the number of actual replicators present). With the right fitness function, evolution can turn rule sets that are sterile into ones that are fecund; the process usually takes 150 or so generations.

Although this technique is limited by the enormous amount of computation involved, the results have been intriguing. Self-replicating structures discovered in this fashion work in a fundamentally different way than self-replicating loops do. For example, they move and deposit copies along the way—unlike replicating loops, which are essentially static. And although these newly discovered replicators are nontrivial in the sense that they consist of multiple, locally interacting components, they do not have an identifiable self-description; that is, there is no obvious genome. The ability to replicate without a self-description may be relevant to questions about how the earliest biological replicants originated. In a sense, researchers are seeing a continuum between nonliving and living structures.

Many researchers have tried other computational models besides the traditional cellular automata. In asynchronous cellular automata, cells are not updated in concert; in nonuniform cellular automata, the rules can vary from cell to cell. Another approach altogether is A. K. Dewdney's Core War and its successors, such as ecologist Thomas S. Ray's Tierra system. In these simulations the "organisms" are computer programs that vie for processor time and memory. Ray has observed the emergence of organisms of various sizes and several forms of "parasites" that co-opt the self-replication code of other organisms.

Getting Real

So what good are these machines? Can they accomplish tasks besides replication? Von Neumann's universal constructor can compute in addition to replicate, but it is an impractical beast. A major advance has been the development of simple yet useful replicators. In 1995 Gianluca Tempesti of the Swiss Federal Institute of Technology simplified the loop self-description so that it could be interlaced with a small program—in this case, one that would write out the acronym of his laboratory, "LSL." His insight was to create automata rules that allow loops to replicate in two stages. First the loop, like Langton's loop, makes a copy of itself. Once finished, the daughter loop sends a signal back to its parent, at which point the parent sends the instructions for writing out the letters.

Drawing letters was just a demonstration. The following year Jean-Yves Perrier, Jacques Zahnd and one of us (Sipper) designed a self-replicating loop with universal computational capabilities—that is, with the computational power of a universal Turing machine, a highly simplified but fully capable computer. This loop has two "tapes," or long strings of components, one for the program and the other for data. The loops can execute an arbitrary program in addition to self-replicating. In a sense, they are as complex as the computer that simulates them. Their main limitation is that the program is copied unchanged from parent to child, so that all replicated loops carry out the same set of instructions.

In 1998 Hui-Hsien Chou and Reggia swept away this limitation. They showed how self-replicating loops carrying distinct information, rather than a cloned program, can be used to solve a problem known as satisfiability. The loops can determine whether the variables in a logical expression can be assigned values such that the entire expression evaluates to true. This problem is NP-complete—in other words, it belongs to the family of nasty puzzles, including the famous traveling-salesman problem, for which there is no known efficient solution. In Chou and Reggia's cellular-automata universe, each replicator received a different partial solution. During replication, the solutions mutated, and replicators with promising solutions were allowed to proliferate while those with failed solutions died out. The loops provide a new paradigm for designing a parallel computer, either from transistors or from chemicals.

Although various teams have created cellular automata in electronics hardware, such systems are probably too wasteful for practical applications; cellular automata were never really intended to be implemented directly. Their purpose is to illuminate the underlying principles of replication and, by doing so, inspire more concrete efforts. In 1980 a NASA team led by Robert Freitas, Jr., proposed planting a factory on the moon that would replicate itself, using local lunar materials, to populate a large area exponentially. Indeed, a similar probe could colonize the entire galaxy, as physicist Frank J. Tipler of Tulane University has argued. In the nearer term, computer scientists and engineers have experimented with the automated design of robots. Although these systems are not truly self-replicating—the offspring are much simpler than the parent—they are a first step toward fulfilling the queen of Sweden's request.

Should physical self-replicating machines become practical, they and related technologies—such as genetics and atomic-scale nanomachines—will raise difficult issues, including the *Terminator* film scenario in which artificial creatures outcompete natural ones. We prefer the more optimistic, and more probable, scenario that replicators will be harnessed to the benefit of humanity. The key will be taking the advice of the 14th-century English philosopher William of Ockham: *entia non sunt multiplicanda praeter necessitatem*—entities are not to be multiplied beyond necessity.

Until August, 2001, the U.S. Government's position on stem cell research was foggy. The Clinton administration allowed the research, but any lab that received federal funding was barred from using that money to pursue stem cell research.

The Bush administration has decided to support stem cell research with federal funds—with some restrictions. In the text from his speech, which follows, President Bush outlines his approach to the sensitive ethical issues surrounding the issue and stresses the pragmatic need that gave rise to his decision.

Remarks by the President on Stem Cell Research

President George W. Bush

THE WHITE HOUSE
Office of the Press Secretary

For Immediate Release August 9, 2001
REMARKS BY THE PRESIDENT
ON STEM CELL RESEARCH
The Bush Ranch
Crawford, Texas

8:01 P.M. CDT

THE PRESIDENT: Good evening. I appreciate you giving me a few minutes of your time tonight so I can discuss with you a complex and difficult issue, an issue that is one of the most profound of our time.

The issue of research involving stem cells derived from human embryos is increasingly the subject of a national debate and dinner table discussions. The issue is confronted every day in laboratories as scientists ponder the ethical ramifications of their work. It is agonized over by parents and many couples as they try to have children, or to save children already born.

The issue is debated within the church, with people of different faiths, even many of the same faith coming to different conclusions. Many people are finding that the more they know about stem cell research, the less certain they are about the right ethical and moral conclusions.

My administration must decide whether to allow federal funds, your tax dollars, to be used for scientific research on stem cells derived from human embryos. A large number of these embryos already exist. They are the product of a process called in vitro fertilization, which helps so many couples conceive children. When doctors match sperm and egg to create life outside the womb, they usually produce more embryos than are planted in the mother. Once a couple successfully has children, or if they are unsuccessful, the additional embryos remain frozen in laboratories.

Some will not survive during long storage; others are destroyed. A number have been donated to science and used to create privately funded stem cell lines. And a few have been implanted in an adoptive mother and born, and are today healthy children.

Based on preliminary work that has been privately funded, scientists believe further research using stem cells offers great promise that could help improve the lives of those who suffer from many terrible diseases—from juvenile diabetes to Alzheimer's, from Parkinson's to spinal cord injuries. And while scientists admit they are not yet certain, they believe stem cells derived from embryos have unique potential.

You should also know that stem cells can be derived from sources other than embryos— from adult cells, from umbilical cords that are discarded after babies are born, from human placenta. And many scientists feel research on these type of stem cells is also promising. Many patients suffering from a range of diseases are already being helped with treatments developed from adult stem cells.

However, most scientists, at least today, believe that research on embryonic stem cells offer the most promise because these cells have the potential to develop in all of the tissues in the body.

Scientists further believe that rapid progress in this research will come only with federal funds. Federal dollars help attract the best and brightest scientists. They ensure new discoveries are widely shared at the largest number of research facilities and that the research is directed toward the greatest public good.

The United States has a long and proud record of leading the world toward advances in science and medicine that improve human life. And the United States has a long and proud record of upholding the highest standards of ethics as we expand the limits of science and knowledge. Research on embryonic stem cells raises profound ethical questions, because extracting the stem cell destroys the embryo, and thus destroys its potential for life. Like a snowflake, each of these embryos is unique, with the unique genetic potential of an individual human being.

As I thought through this issue, I kept returning to two fundamental questions: First, are these frozen embryos human life, and therefore, something precious to be protected? And second, if they're going to be destroyed anyway, shouldn't they be used for a greater good, for research that has the potential to save and improve other lives?

I've asked those questions and others of scientists, scholars, bioethicists, religious leaders, doctors, researchers, members of Congress, my Cabinet, and my friends. I have read heartfelt letters from many Americans. I have given this issue a great deal of thought, prayer and considerable reflection. And I have found widespread disagreement.

On the first issue, are these embryos human life—well, one researcher told me he believes this five-day-old cluster of cells is not an embryo, not yet an individual, but a pre-embryo. He argued that it has the potential for life, but it is not a life because it cannot develop on its own.

An ethicist dismissed that as a callous attempt at rationalization. Make no mistake, he told me, that cluster of cells is the same way you and I, and all the rest of us, started our lives. One goes with a heavy heart if we use these, he said, because we are dealing with the seeds of the next generation.

And to the other crucial question, if these are going to be destroyed anyway, why not use them for good purpose— I also found different answers. Many argue these embryos are byproducts of a process that helps create life, and we should allow couples to donate them to science so they can be used for good purpose instead of wasting their potential. Others will argue there's no such thing as excess life, and the fact that a living being is going to die does not justify experimenting on it or exploiting it as a natural resource.

At its core, this issue forces us to confront fundamental questions about the beginnings of life and the ends of science. It lies at a difficult moral intersection, juxtaposing the need to protect life in all its phases with the prospect of saving and improving life in all its stages.

As the discoveries of modern science create tremendous hope, they also lay vast ethical mine fields. As the genius of science extends the horizons of what we can do, we increasingly confront complex questions about what we should do. We have arrived at that brave new world that seemed so distant in 1932, when Aldous Huxley wrote about human beings created in test tubes in what he called a "hatchery."

In recent weeks, we learned that scientists have created human embryos in test tubes solely to experiment on them. This is deeply troubling, and a warning sign that should prompt all of us to think through these issues very carefully.

Embryonic stem cell research is at the leading edge of a series of moral hazards. The initial stem cell researcher was at first reluctant to begin his research, fearing it might be used for human cloning. Scientists have already cloned a sheep.

Researchers are telling us the next step could be to clone human beings to create individual designer stem cells, essentially to grow another you, to be available in case you need another heart or lung or liver.

I strongly oppose human cloning, as do most Americans. We recoil at the idea of growing human beings for spare body parts, or creating life for our convenience. And while we must devote enormous energy to conquering disease, it is equally important that we pay attention to the moral concerns raised by the new frontier of human embryo stem cell research. Even the most noble ends do not justify any means.

My position on these issues is shaped by deeply held beliefs. I'm a strong supporter of science and technology, and believe they have the potential for incredible good—to improve lives, to save life, to conquer disease. Research offers hope that millions of our loved ones may be cured of a disease and rid of their suffering. I have friends whose children suffer from juvenile diabetes. Nancy Reagan has written me about President Reagan's struggle with Alzheimer's. My own family has confronted the tragedy of childhood leukemia. And, like all Americans, I have great hope for cures.

I also believe human life is a sacred gift from our Creator. I worry about a culture that devalues life, and believe as your President I have an important obligation to foster and encourage respect for life in America and throughout the world. And while we're all hopeful about the potential of this research, no one can be certain that the science will live up to the hope it has generated.

Eight years ago, scientists believed fetal tissue research offered great hope for cures and treatments—yet, the progress to date has not lived up to its initial expectations. Embryonic stem cell research offers both great promise and great peril. So I have decided we must proceed with great care.

As a result of private research, more than 60 genetically diverse stem cell lines already exist. They were created from embryos that have already been destroyed, and they have the ability to regenerate themselves indefinitely, creating ongoing opportunities for research. I have concluded that we should allow federal funds to be used for research on these existing stem cell lines, where the life and death decision has already been made.

Leading scientists tell me research on these 60 lines has great promise that could lead to breakthrough therapies and cures. This allows us to explore the promise and potential of stem cell research without crossing a fundamental moral line, by providing taxpayer funding that would sanction or encourage further destruction of human embryos that have at least the potential for life.

I also believe that great scientific progress can be made through aggressive federal funding of research on umbilical cord placenta, adult and animal stem cells which do not involve the same moral dilemma. This year, your government will spend $250 million on this important research.

I will also name a President's council to monitor stem cell research, to recommend appropriate guidelines and regulations, and to consider all of the medical and ethical ramifications of biomedical innovation. This council will consist of leading scientists, doctors, ethicists, lawyers, theologians and others, and will be chaired by Dr. Leon Kass, a leading biomedical ethicist from the University of Chicago.

This council will keep us apprised of new developments and give our nation a forum to continue to discuss and evaluate these important issues. As we go forward, I hope we will always be guided by both intellect and heart, by both our capabilities and our conscience.

I have made this decision with great care, and I pray it is the right one.

Thank you for listening. Good night, and God bless America.
END 8:12 P.M. CDT

hburgh reported in February
y had cloned an adult sheep
ediately captivated the pub
some cautious researchers
ther the success might not
the pan. After all, the Sco
kers had to try 277 times b
ceeded in producing the clo
ly. Unless the efficiency o
hing process could be great

Conclusion

Cloning will not raise Adolf Hitler from the dead. Genghis Khan will never walk again, and a lock of Napoleon's hair harbors no hope for putting him back into European politics.

No matter how technically advanced cloning may become in the years ahead, identical is never exactly identical. Why? Some scientists believe it is these tiniest of shifts that take place on a microscopic level that activate some traits while others remain latent. Others believe that environmental and circumstantial choices are all that separate the power tyrant from the painter.

As long as cloning remains such a potent science, it will continue to attract the attention of a surprised public, not really sure what to make of transgenic animals or the thought of a lesbian couple conceiving a child through cloning. Many other legitimate questions come up in the wake of the fledgling science rushing to achieve. To move forward in a meaningful way, the science community needs to address all questions, not only to put the public more at ease, but to clearly define for themselves where they intend to go.

Aborted fetuses, as stem cell source, 73

Accommodation, in xenotransplantation, 82

Adenine-based replicators, 106

Adenine ribose naphthalene imide (ARNI), 103
  as self-replicating molecule, 112

Adenine ribose thymine (ART), 106

Adenines, 103, 104

Adult animals, cloning of, 35, 37–39, 47, 50, 51–53

Adult stem cells, 70–71

Advanced Cell Technology (ACT), 26
  cloning of bucardo by, 34
  cloning of giant panda by, 30
  cloning of livestock by, 43–45
  cloning of pets by, 32–33

Aging
  of cloned sheep, 90
  stem cells versus, 70

Agriculture
  asexual plant propagation in, 9
  cloning in, 5
  increasing crop yield in, 11–12
  potato blight in, 11

AIDS, 52, 59

Allergic reactions, to antibodies, 17–18

Alzheimer's disease, 80, 90, 127

Amino acids, 62, 100

Amyotrophic lateral sclerosis, 71, 86

Animals
  antibodies from cloned cells of, 17–18
  cloning of, 1–3, 89
  cloning of adult, 37, 38–42, 44, 47, 50
  cloning of endangered, 21, 23–26, 26–27, 27–34, 34–35
  genetically altered, 41, 43–45
  mother, 25
  sperm banks of endangered, 34–35
  transgenic, 2, 43, 46-47, 57
  in xenotransplantation, 80–81

Antibiotic-resistance genes, 49

Antibiotics, from plants, 17–19

Antibodies. *See also* Plantibodies
  allergic reactions to, 17–18
  in transplant rejection, 81

Anticlotting proteins, transgenically produced, 65

Artificial insemination, of pandas, 27.
  See also In vitro fertilization

Asexual reproduction
  cloning as, 7–8, 8–9
  of Russet Burbank potato, 10–11

Audubon Institute Center for Research of Endangered Species (AICRES), 26–27, 34
  cloning of bongo antelope by, 30–31
  cloning of pets and, 32, 33

Baboons, xenotransplantation into, 84–85, 85

Bi-cycles, of self-replicating molecules, 103, 104

Bioreactors, 60
  transgenic animals as, 57

Blastocysts, 26, 70

Blight, of potato, 11
Blind Watchmaker, The (Dawkins), 100
Blood
    infectious agents in donated, 59
    in transplant rejection, 81, 83, 85
    working with stem cells in, 76
Blood proteins
    isolation of, 61
    from transgenic animals, 57-66
Body cells, in cloning, 7–8
Bone marrow
    stem cells and, 70-71
    in xenotransplantation tolerance, 84
Bongo antelope, cloning of, 24, 27, 30–31
Bovine spongiform encephalitis (BSE), 51
Brain, stem cells in repairing damaged, 71
Bucardo mountain goat, cloning of, 24, 31–34
Burn victims, self-cloning in treatment of, 89
Bush, George W., 121
    on human stem cell research, 123-128
Cancer, 17, 71
Capillary destruction, in transplant rejection, 81
Carbohydrates, blocking of plant-produced human antibodies, 19
Cell membranes, life processes and, 107
Cells. *See also* Donor cells; Egg cells; Somatic cells; Stem cells
    from cloned embryos, 51
    in cloning, 1–2, 7–8, 45-47
    cryopreservation of, 24–25
    encapsulated, 86
    immunoisolation of, 86
    totipotency of, 9
    in transgenic animals, 49
    universal human donor, 51
Cellular automata
    emergent self-replication, 113
    practical, 118
    as self-reproducing machines, 111, 112-114
Children
    lifestyle of first cloned, 96
    mass producing, 91
Chromosomes

in animal cloning, 21, 28–29, 34, 40, 50
    of cloned sheep, 90
    mutations and, 93
    of quiescent cells, 54
    of tobacco, 10
    of transgenic mice, 60
Complementary molecules, self-replicating, 106-107
Complexes, of self-replicating molecules, 106-107
Complexity, of self-reproducing machines, 114
Computers, self-reproducing machines simulated in, 109, 111-113
Convention on International Trade in Endangered Species (CITES), 23
Corn, human antibodies from, 17, 18–19
Creutzfeldt-Jakob disease, 59
Cryopreservation, of cells, 24–25
Cystic fibrosis, 50, 94
Cytoplasm, in cloning, 25–26, 28–29
Differentiation, of stem cells, 52
Diabetes
    encapsulated cells in treating, 86
    human stem cells treating, 52, 76
    islet replacement in treating, 81
    pig insulin for treating, 65
    self-cloning in treatment of, 90
    transgenic animals treating, 50
    treating via xenotransplantation, 80
Diaminotriazine xanthene biphenyl imide (DIXBI), recombination of, 106
Diaminotriazine xanthene thymine (DIXT), in recombination, 106
Differentiation, of stem cells, 52, 70, 73, 90-91
Disease resistance, cloning for, 94-95
DNA
    in cloning, 2, 3, 8, 47, 50-51, 53
    as loop, 117
    of mammoths, 31
    mutations and, 105
    origin of life and, 101
    as replicating molecule, 106
    replication of, 101
    structure of, 101

of thylacine, 31
of transgenic animals, 52-53,
59-62, 64
Dogs
cloning of, 32–33
reproductive physiology of, 33
in xenotransplantation, 80
Dolly (sheep), cloning of, 2, 35,
38-40, 41, 43, 45, 47, 49, 50,
90, 91
Donated blood, infectious agents, 59
Donor cells, 81
in animal cloning, 43
in nuclear-transfer experiments,
53
universal human, 51
Drugs
immunosuppressive, 84
from plants, 17–19
from transgenic animals, 2, 57–66
Earth, origin of life on, 97, 100-102
Egg cells
in animal cloning, 21, 25–26,
28–29, 46, 49
in gene therapy, 55
in human cloning, 90-91
of transgenic animals, 49
Electric pulses, in cloning, 25–26, 29,
44, 47
Embryonic fibroblast cells, 2
Embryos
in animal cloning, 45
creation of, 126
as human life, 125-126
human stem cells from, 52,
124-129
stem cells from, 69, 73, 74
stem cells from cloned, 52
storage of, 138, 139
Endangered animals
cloning of, 21, 23–26, 26–27,
27–34, 34–35
and cloning of pets, 33
sperm banks of, 34–35
Ethics
of animal experimentation, 66
in biosciences, 46
of creating transgenic animals, 50
of genetic engineering, 39, 66
of human cloning, 41, 50, 52,
89, 95
of human stem cell research,
52, 69, 70-73, 122, 124–128
of nuclear transfer process, 74

Eugenics, cloning for, 94
Evolution
cellular automata simulation of,
117-118
mutations and, 105
of self-replication, 111
Extinct animals, resurrection via
cloning of, 21, 34
Factor IX, 49, 53
transgenic production of, 58
Federal funding, stem cell research
and, 73, 122, 124-129
Fibroblasts, 29, 47
Fitness function, 117-118
Fluorescent tags, 94
Foreign tissue
immunoisolation of, 86
rejection of, 77
Furin, protein C and, 63
Gaur, 24
cloning of, 23–24, 25–27
Gene pools, cloning and, 24–25
Genes
of cloned children of lesbian
couples, 92
cloning and, 8, 24–25
in fusing protoplasts, 13
human personality and, 53
origin of life and, 102
in sexual reproduction, 7
transgenic animals and, 2, 49,
54, 57-66
Gene therapy, 51
human cloning in, 94
Genetically altered animals, 41,
43–45
Genetic code, 62, 93
Genetic diseases, cloning to avoid
transmitting harmful, 92
Genetic engineering, ethics of, 37
Genetic manipulation, 61
Genie (pig), as transgenic animal, 66,
58, 62, 63-64, 66
Genomes, of clones, 8–9
Giant panda, cloning of, 24, 27–30
Goats, transgenic proteins from, 65
Habitat preservation, cloning and, 35
Heart cells, turning stem cells, 75
Hemophilia, 49
treating via xenotransplantation,
80
treatment of, 58
Hodgkin's lymphoma, therapeutic
antibodies for, 18

House cat
in African wildcat cloning, 27
cloning of, 32–33
Human body, stem cells in
understanding of, 75
Human cloning, vii–viii, 2, 34, 89,
90–96
ethics of, 39, 39, 50, 52, 89,
95
in gene therapy, 94
identity and, 94
for lesbians, 92
mass production of humans, 91
nightmare scenarios of, 91
treating infertility via, 91
ultimate uses of, 95-96
Human donor cells, universal, 51
Human factor IX, 50, 59. *See also*
Blood proteins
transgenic production of, 61
Human factor VIII, 59
Human gene pool, cloning and,
92
Human personality, genes and, 53
Humans
animals endangered by, 35
embryos as, 125-126
harvesting organs from cloned, 95
therapeutic animal proteins in, 65
transplanting genetically modified
organs into, 49-50
in xenotransplantation, 80
xenotransplantation tolerance in,
81-87
Human stem cells, 2-53. *See also*
Stem cells
ethical issues concerning, 52-53
isolation of, 71
research with, 122, 124-128
Hydrogen bonds, in self-replicating
molecules, 102, 104-105
Hyperacute immune reactions, 50, 77,
81
overcoming, 82–83, 83
Identical twins
clones as, 91, 96
in kidney transplants, 80–81
Identity, cloning and, 94–96
Imides, 101, 104–105
Immortality, of stem cells, 74–75
Immune system
organ transplants and, 50
in overcoming *hyperacute* immune
reactions, 82–83, 83

stem cell rejection and, 74
xenotransplantation and, 80–81
in xenotransplantation tolerance,
83–87, 85
Immunoisolation, 85–87
Immunosuppressive drugs, 84
Infectious agents, donated blood, 60
Infertility, human cloning assist, 91
Infertility treatments, 73
Infinite regress, in describing self-
reproducing machines, 111–112
Instructions, in self-reproducing
machines, 115–116
Insulin
from pigs, 65
transplanted pancreatic cells, 81
Intracytoplasmic sperm injection, in
treating male infertility, 91
In vitro fertilization, 25
first human born via, 90, 91
Islets, pancreatic, 81, 86–87
Kemp's triacid, in molecular
replication, 104
Kidney transplants, first successful,
80–81
Killer T cells, in xenotransplantation
tolerance, 85
Leaf-cell protoplasts, cloning from, 5,
7–13
Leaves, small terminal, 5
Lesbians, cloned children for,
91-92
Leukemia, 71, 127
Life
cell boundaries and, 106-107
embryos as human, 125-126
origin of, 97, 100-102
sacredness of, 52, 127
synthetic self-replicating
molecules and, 107
Livestock
cloning to improve, 43–45, 49,
51–53
transgenic, 46, 51, 57–66
Loops, in self-reproducing machines,
115–116, 116-117, 118, 120
Lou Gehrig's disease, 71, 86
Machines
cloning of, 109, 111-118
self-replicating, vii
Mammals
cloning of, 37, 37–39, 49
medical uses of transgenic, 57–66
Mammary glands, 49

activating foreign genes in mouse, 61, 63–64
as transgenic protein producers, 65–66
Mammoths, cloning of, 21, 31
Membranes, life process and, 106–107
Messenger RNA, in cloning, 46–47, 53–54
Mice
    stem cells and new neurons in, 71
    transgenic, 60–62
    in treating cystic fibrosis, 50
Microinjection, transgenic animals via, 49
Milk, human proteins in animal, 49, 57, 61-62, 63–64, 65–66,
Mother animals, in cloning, 25, 25–26, 61, 62
Mouse cells, antibodies from cloned, 17–18
Mouse milk protein, in transgenic mice, 61, 64
Mouse promoter, in transgenic animals, 62
Multiplicity, 91
Muscular dystrophy
    eliminating, 94
    human stem cells in treating, 52
    transgenic animals in treating, 50-51
Mutations
    in cellular automata, 116-117
    cloning to avoid transmitting harmful, 92–93
    evolution and, 105
Nanotechnology, 111
Naphthalene, 103
Neurons, growing new, 71–72
NP-complete problems, 118
Nuclear transfer process
    in cloning, 25–26, 28–29, 34, 46-48, 53–54, 90, 91
    ethics of, 74
Nuclei, in cloning, 1–2, 28–29, 38, 44, 45, 46
Nucleic acids, recombination of, 106
Old-age dementia, human stem cells in treating, 72
Organ replacements
    self-cloning for, 90
    through stem cells, 70–71, 73
Organs
    genetically modified, 49–50

harvesting from cloned humans, 95
immunoisolation of, 86
shortage of donated, 81
Organ transplants. *See also* Xenotransplantation
    with genetically modified organs, 49–50
    rejection of, 77
    stem cells and, 73, 75
Pancreas, 81. *See also* Islets
Pancreatic cancer, 74
Panda, cloning of, 24, 28–30
Parallel computers, cellular automata and, 118-119
Parkinson's disease
    human stem cells in treating, 52, 71–72
    self-cloning in treatment of, 91
    transgenic animals in treating, 50–51
    treating in rats, 70
    treating via xenotransplantation, 80
Patents, for cloning, 47
Pathogen-free animals, 65
Pathogens
    in donated blood, 60
    from transgenic animals, 65, 66
Perfused devices, in immunoisolation, 86–87
Pets, cloning of, 32–33
Phytophthora infestans, 11
Pigs
    in immunoisolation tests, 86–87
    insulin from, 65
    organ transplants from, 50, 79–80, 82, 83, 84–85, 85
    See also Xenotransplantation
    as transgenic animals, 57, 58, 62, 82
Plantibodies, 17–19
Plants
    cloning of, 1, 5, 7–13, 89
    human antibiotics from, 17–19
    medicines from, 15, 17–19
    varieties of, 9–10
    vegetative mutations of, 9
Plastics, in immunoisolation tests, 86–87
Polar bodies, 28–29
Polar interactions, among self-replicating molecules, 102

Post-translational modifications, to protein molecules, 62–63
Potato, Russet Burbank, 10–11, 12–13
Potato blight, 11
Potato plants, cloning of, 7–13
Pre-embryos, 125-126
Primordial soup, for cellular automata, 115
Prions, in bovine spongiform encephalitis, 51
Protein C, 58
 furin and, 63–64
 transgenic production of, 62
Protein molecules
 amino acids in, 62
 post-translational modifications to, 62–63
Protein shells, life process, 106-107
Protein synthesis, 62–63
Protoclones
 potato, 11–13
 tobacco, 10
 yield advantages of, 12
Protoplast cloning, 7–13
 defined, 8
Protoplasts
 fusing, 13
 potato, 11–12
 tobacco, 9–10
Quiescence, in cloning, 47, 53–54
Rabbits, 30, 60, 80
Radiation therapy, xenotrans-plantation tolerance via, 84–85
Randomness, 116-117
Recombination
 of DNA, 105
 of self-replicating molecules, 105-107
Regeneration, of potato plants, 7–13
Rejection
 antibodies in, 81
 of donor organs, 77
 of foreign stem cells, 74
 in kidney transplants, 80–81
Relative position measure, for fitness function, 118
Repository network, for cloning, 34
Resistance
 to potato blight, 11
 through fusing protoplasts, 13
Ribose, 90–102
RNA, origin of life and, 90–102
Roslin Institute, 37, 38, 43, 45, 47,

49, 90
Russet Burbank potato, 10–11, 12–13
Satisfiability, cellular automata for solving, 118
Self-complementary molecules, 103-104
Selfish Gene, The (Dawkins), 101
Self-repairing silicon chips, 111
Self-replicating factories, 118
Self-replicating molecules, 2–3, 97, 100–107
Self-reproducing machines, 109, 111—118
 computer simulation of, 100-104
 design of, 99–100, 116–117
 nontrivial, 114–116
 practicality of, 118
 spontaneous emergence, 116-117
Sexual reproduction, 7
 human cloning versus, 92-93, 95-96
Sheep
 cloning of, 2, 35, 37–39, 41, 43, 45, 46–47
 transgenic, 49
 in treating cystic fibrosis, 50
Sheep milk protein, 59–60
Side effects, of transgenic proteins, 66
Sigmoidal growth, 104–105
Skin cells, in cloning, 28–29
Solvents, for self-replicating molecules, 102–103
Somatic-cell nuclear transfer, 90, 91. *See also* Nuclear transfer process
Somatic cells, in cloning, 7–8
Sperm banks, of endangered animals, 34–35
Sperm donors, for lesbian couples, 91-92.
Spinal cord injuries, 90–92
Sports, 9, 10–11
Stem cells, 69, 70. *See also* Human stem cells as body repair mechanism, 70–75
 cloning of, 2, 69
 commercial therapies with, 74–75
 creating human, 52–53
 differentiation, 52, 70, 72–73
 ethical issues concerning, 52–53
 funding research on, 82–83
 future promise of, 74–75
 immortality of, 72–73

isolation of, 51, 71
nuclear transfer process with, 72
organ replacements via, 70–71
rejection of foreign, 72
Stem cell techniques, availability, 73
Surrogate mothers, 91
Synthetic molecules, self-replicating, 2–3, 97, 100–107
Therapeutic antibodies, 17–19
Thymine-based replicators, recombination of, 106
Tissue plasminogen activator, 58, 61
Tissues
immunoisolation of, 86
self-cloning to replace, 90–91
Tobacco
cloning of, 5, 9–10
human enzymes from, 19
protoclones of, 10
Tolerance, for xenotransplantation, 83–87
Transgenic animals, 2, 49–51, 55
ethics of creating, 50
ethics of using, 66
human blood proteins from, 57–60, 61-66
initial experiments with, 60–61

Transplants. *See also* Xenotransplantation
genetically modified organs,49–50
stem cells and, 74, 75
Twins
clones as identical, 91, 96
in kidney transplants, 80–81
Umbilical blood, working with stem cells in, 74
U.S. Department of Agriculture, 8, 45
Universal constructor, 114, 115, 118
Universal human donor cells, 51
Viruses, life process and, 106–107
Von Neumann, John, 111–113, 115, 116-117
Whey acidic protein, 62
Xenotransplantation, 79–87
early attempts at, 80–81
hyperacute immune reactions and, 82–83
tolerance for, 83–87
Zona pellucida, 25, 28–29
Zoos, cloning in, 24–25

Photo Credits

Page 24 Roland Seitre, Peter Arnold Inc. Page 28-29 Laurie Grace (Illustrations); Philip Damiani (photographs). Page 83 Jennifer C. Christiansen. Page 85 Richard Jones and Jennifer C. Christiansen. Page 102 & 104 Jared Schneidman/ JDS.